DESTINY: A FUTURE BEYOND YOUR WILDEST DREAMS

An original story about the future and destiny of the human species!

I dedicate this book to my Mom, Dad and little Sister as well as my closest friends and family who have allowed me to have the many fun experiences, quality of life and helped me achieve so much in life without whom writing this book would be next to impossible!

MYSELF

I'm Nitish Kannan, and I grew up on the space Coast in Cape Canaveral, Florida. I think this city is a paragon of excellence, from which humanity launched a man to the moon and remains the metaphorical home of the American space program. But, to me it has always been so much more. I've been fortunate to meet a few astronauts and on occasion listen to them speak. Often times, when I recall what I hear astronauts whom have visited space talk about and describe what they see and share their experience, I always hear the same thing. I hear them lament, I see no countries with borders, and that it's hard to imagine or even fathom why anyone would be fighting over the borders of nations, religions and resources, when clearly the whole world is large, from space there are no visible borders and there is enough resources for all. When clearly we have more in common than that separates us. When clearly our DNA illustrates we have more in common than that separates us. The Earth is so small in the context of the universe and we get 10000x more solar power than we need in the form of usable photons which can become electricity. We have oceans with water, maybe they have little fishies and a bit of salt in them. However, with the advent of cheap power thanks to nuclear fusion and solar we can have cheap water. I don't see resource shortages in our universe they say. The galaxy is large and nuclear fusion is an infinitely renewable process all over the universe.

So why don't people look at the world like astronauts? They look at the world with astonishment, thankful for its large scale and our uniqueness and have a sense of wonder. Why don't we change our conversations to wonder and how much better it can be rather than complain about things that aren't real, false prophets, false ideology and false wars haven't done much over the last few thousand years, so why bother dealing with them now. There is our universe and there are no laws on humanity. So be free, think big and change the world. The biggest change in your life starts with you.

So when people tell me heads in the sky or I'm dreaming, I tell them that's the problem, you're heads not looking at problems from the sky. Instead you're looking at problems from a not so ugly place down there. That's why at the age of 12, I became fascinated and obsessed with the future, I understood at 12 computer power was doubling yearly, technology was getting fast, the internet was just taking off. I felt like the kids might has in the 60's when the moon program was full swing, except in the technology world. I figured only the future mattered and nothing else did! I say to myself often, you can't change the past but you can make your own future! Therefore I wanted to contribute and predict how technology would make life better not just for me, but for a billion people on Earth. I wrote this book to offer my foresight and tell people that the future is getting better, technology will make life better and that if you're prepared with an inside knowledge on what's coming you shall reap the benefits and be ready for our amazing future!

CONTENT'S

Preface

Hero's

SAGA I

IT'S THE ECONOMY STUPID

SAGA II

TECHNOLOGY EVOLUTION

SAGA III

POWERING OUR HUMAN MACHINE FUTURE

SAGA IV

LET'S GO GALACTIC

Preface

You know that curious kid in you that dreamed about a future of possibility, well that's not far away at all and in fact it's happening a lot faster than you think, I shall show you what's coming and how to prepare. After all, the best way to predict the future is to make it yourself and we're certainly working harder than you can imagine!

This book is supposed to scare you right? If reading this doesn't frighten you, then I haven't done my job. The most important thing in this book that I want my friends to take away is how our entire civilization is predicated on accelerating technological change. It means that all technological progress is the backbone of human civilization improving. As the prominent venture capitalist and first investor in Facebook, Peter Thiel says, our entire civilization has grown because of vast improvements in technology. The worlds most respected futurist Ray Kurzweil says, we didn't say on the ground, we didn't stay on the planet and we won't stay within the limits of our biology. In less than 40 years we will become like the Gods with magical technologies that would blow your mind today, as my friend Jason Silva says, "Having created the Gods, we shall become like the GODS!"

This century will make the biggest shift in our humanity from a type zero to a type 1 civilization (explained later in the book) and we will solve many of the grand challenges of the age old, like curing aging, have mind uploading and unlimited clean and free energy. We shall even create new technologies that will make science fiction into science fact.

You do want to make it to the future right? I hope this book motivates you to be aware at the possibility of what's coming. The new digital renaissance that's coming soon. Very soon at hand. I mean we are going to be living in times of great prosperity and technological marvel, the 21st century will see 20000 years of progress due to the exponential nature of technology. Not sure? Well I shall explore futurists Michio Kaku and Raymond Kurzweil's claims who believe so. I firmly believe that the greatest technological advances will take place over this century. I will examine the credibility of new technological advances in a fun way! Well read on and find out!

In this book I will tell you about how the future will change for the better and how you can capitalize on that growth to become a better inventor and better apply this knowledge to enhance your life.

We are going to be living in very prosperous times soon, In fact I would call it a modern Renaissance, a time when our basic needs for every man, woman and child are met so that we can do greater things that our higher intellect is reserved for, like expanding discoveries in science, making movies and entertainment we enjoy, composing new songs, and inventing new technologies which will even more enhance our humanity.

In fact, we already have autonomous cars like the Google cars, new humanoids that can clean our house, and artificial intelligence assistants like IBM's Watson and Siri that can access all of human knowledge and respond with intelligent results. I foresee a time in the next 5-10 years we shall reap the benefits of these advances and machines and we'll have less wasteful work to do!

I make a bold prediction in this book that by mid-century we will reach a point that many have called the technological singularity, a phrase coined by Vernor Vinge, and popularized by the prominent futurist Ray Kurzweil. As absurd as my predictions may sound they're quite conservative and appear quite in line with some of the greatest minds of our time. The technological singularity is not one event but the culmination of thousands of events and advances of our technology.

I think you should plan to stick around, stay healthy and experience the benefits of the technology we are inventing, because if you thought the world is getting worse I certainly have the facts to convince you otherwise. I make the bold prediction that if you can hang around 20 more years, at which point, the chart on Moore's law will allow for intelligent robotic nanobots to cure many diseases, then you will reach the singularity by mid-century where we can augment our bodies with new robotic nanobots to replace our brains and organs and merge with machines. In fact I shall take a look at tomorrow's robots, full immersion virtual reality 3D environments and even mind reading machines! As futuristic as this may sound, this is actually far from science fiction, in fact over the last 30 years many of the exponential progresses needed for the technological singularity are right on track, there is little reason to doubt they will not continue, in fact yearly, we're seeing exponential progress of technology measured utilizing many metrics follow the trends pointed out by Ray Kurzweil and many other futurists continue right on track. In fact many people point out that there will great a great disparity between the have and have not's right of the back when I even mention technological progress, however the most important fact to remember throughout this book is that technology doubles in power every 2 years and deflates in value by half. So you get twice the power for half the price, no matter how you measure it. In fact many prominent people have gone on to say that the rapid growth of information technology and the drastic deflation of technology is the primary force behind keeping rampant economic inflation from happening. Thus, if you're convinced the world is getting poorer, I hope I can convince you technology is actually making life better for all!

I would like to thank the many heroes who have made amazing accomplishments and inspired me to write this book, here are a few of them I've met that are technology geeks and proponents of this future!

MY HERO'S

Here is Raymond Kurzweil and I! The famous inventor and personal hero of mine, whom I've given much credit to in this book! Thanks for hanging out with me at CES and bantering on voice recognition, products and technology and inspiring us to stay healthy!

Steve Wozniak and myself, the kindest geek you've ever met. Woz created the Apple II computer and co-founded Apple. He's a 60 something year old man with the heart of a 16 year old kid and thinker! Thank you so much for always being kind and bantering on technology with myself.

My childhood hero who's one of my favorite futurists, fellow science fiction geek and Theoretical physicist. Dr. Michio Kaku, who so kindly

signed my copy of his book "Physics of the future."

SAGA 1

"It's the economy stupid"

LESSONS OF AN ECONOMIST FOR THE CONDITIONS TO BUILD A TECHNOLOGICALLY LITERATE SOCEITY

In this section, we will discuss how to make trillions of dollars in wealth and why it's necessary to start having access to the largest markets in the world. Also, I will discuss how we are empowered to build powerful nations, and not just start companies but startup entire nations of free enterprise! But for that to happen we must understand the conditions that bring about prosperous future.

If anything but my lack of words, I shall start this book with a quote by a friend who gave a great **TED TALK** a few years ago by a great life sciences investor named Juan Enriquez. He starts off with this powerful statement:

"I'm supposed to scare you, because it's about fear, right? Afraid, but You should be really afraid that -- if we stick up the first slide on this thing -- there we go -- that you're missing out. Because if you spend this week thinking about Iraq and thinking about Bush and thinking about the stock market, you're going to miss one of the greatest adventures that we've ever been on. And this is what this adventure's really about. This is crystallized DNA. Every life form on this planet -- every insect, every bacteria, every plant, every animal,

every human, every politician -- (Laughter) is coded in that stuff. And if you want to take a single crystal of DNA, it looks like that. And we're just beginning to understand this stuff. And this is the single most exciting adventure that we have ever been on. It's the single greatest mapping project we've ever been on. If you think that the mapping of America's made a difference, or landing on the moon, or this other stuff, it's the map of ourselves and the map of every plant and every insect and every bacteria that really makes a difference. And it's beginning to tell us a lot about evolution. (Laughter)"

Juan has convinced me that humanity shares more things in common than we have in differences. Also, since we all share life code in the form of DNA, I guess it makes all violence quite barbaric and pointless and a huge waste of our precious society!

So channeling Juan Enriquez's voice, let me ask you; this book is supposed to scare you right? I'm supposed to scare you right? Well I'll try not to and instead tell you it's the media that installs fear into your linear thinking brains. In fact they earn their living on making you glued to the television with negative headlines, the banks are failing, and wars are plaguing the world and so forth. In fact you may see the puppets of the media which appears to be free. You may say all these troubles in the financial market what are they? Well I'll do my best to explain them to you and you will see the fallacy of such theories. Now you may be saying this book is about trends of the future right?

How does economics play into this? Well first of we see that as we cut our budget, we also have to grow. When we look at the Budget of America we see that most of it is taken up with entitlements and social security. Now one important assumption is that government around the world is supposed to take care of us. Well the fallacy of this myth is that government by itself produces no valuable resources on its own. Instead it must steal from productive individuals in society. Moreover it's not fair that government steals from all people in society!

In fact satire is the best medicine to cure an ailing and brainwashed society. I think Sacha Baron Cohen, playing General Aladeen, said it best in his magnum opus movie *The Dictator,* "what if America were a dictatorship?." He yells to the U.N., "Imagine if America was a dictatorship," General Admiral Haffaz Aladeen invites a crowd of wary New Yorkers. He is on a state visit from the North African country he rules with an iron fist — and with a fleet of gold-plated Hummers — and he could not be more sincere.

Source: http://rt.com/art-and-culture/news/tajikistan-bans-dictator-666/

"You could let 1 percent of the people have all the nation's wealth. You could help your rich friends get richer by cutting their taxes and bailing them out when they gamble and lose. You could ignore the needs of the poor for health care and education. Your media would appear free, but would secretly be controlled by one person and his family. ... You could wiretap phones. You could torture foreign prisoners, you could have rigged elections, you could lie about why you go to war, you could fill your prisons with one particular racial group and no one would complain. You could use the media to scare the people to support policies that are against their interests. I know this is hard for you Americans to imagine, but please try!"

If you all didn't get it, he was satirizing American government not differing much from the great dictatorships! Well the Joke is on us!

THE ELEPHANT WILL AWAKEN!

Juan Enriquez in another talk about the financial crisis mentions that the first big bear in the room today is the issue of leverage in our banking system. The problem with leverage is that a commercial bank has 9-10x leverage, whereas an investment bank has 15 times the leverage. What that means is that every dollar put into the bank goes bad over 15 times over when you have 15 times leverage. You know what's even scarier, Bank of America has 21 times leverage, and Citi bank 47 times leverage according to Juan Enriquez. This means every loan by Citi bank goes bad 47 times over. Every dollar you put gets loaned out 47 times which would cripple them and therefore every bank takes donations from tax payers from us. Geez, thanks! According to Enriquez in 1967 we spent, 38 percent on entitlements. In 2017 we may spend 100 percent is entitlements at current rates. Moreover with the money we loan out we're getting into more debt, on top of that every time we have this problem we keep using the Federal Reserve to devalue our currency... We need to banish entitlements and these spending habits. We will end up as Iceland did... we should just default, and allow the market to have a correction. If we do not, then as Japan has been in a 20 inflationary cycle, America will follow a period of stagnation. Moreover if we don't fix this, we'll all end up being trillionares... no really that's what happens when you devalue your currency like many African nations have done. Then all bets are off. We'll lose the dollar. I say the only real solution is to have a currency backed by the gold standard and keep its value. The Nitish Kannan solution? I say eliminate the Federal Reserve System which is the primary cause of these bubbles.

Here is something the mainstream media didn't cover, instead of printing money out of thin air Iceland let its banks fail and the market to work again by liquidating this bad debt, thus their economy is

booming again!

Source: https://www.facebook.com/TheOther98

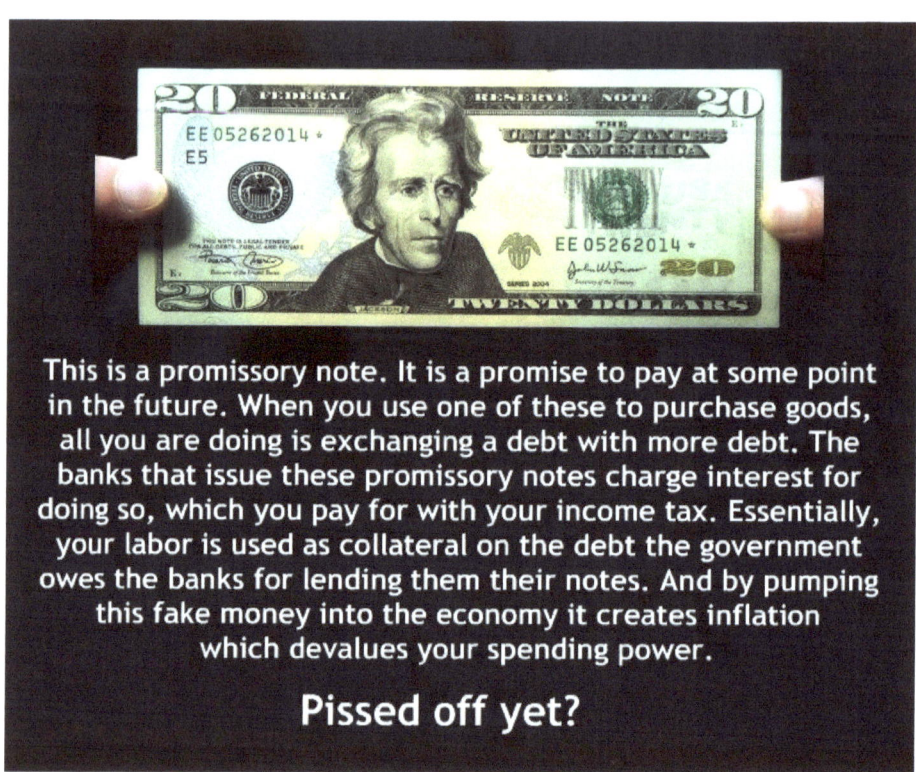

Source: https://www.facebook.com/TheOther98

The next question I ask is what would happen if the Federal Reserve was shut down permanently? That is a question that CNBC asked recently, but unfortunately most Americans don't really think about the Fed much. Most Americans are content with believing that the Federal Reserve is just another stuffy government agency that sets our interest rates and that is watching out for the best interests of the American people. But that is not the case at all. The truth is that the Federal Reserve is a private banking cartel that has been designed to systematically destroy the value of our currency, drain the wealth of the American public and enslave the federal government to perpetually expanding debt. During this election year, the economy is the number one issue that voters are concerned about. But instead of endlessly blaming both political parties, the truth is that most of the blame should be placed at the feet of the Federal Reserve. The Federal Reserve has more power over the performance of the U.S. economy than anyone else does. The Federal Reserve controls the money supply, the Federal Reserve sets the interest rates and the Federal Reserve hands out bailouts to the big banks that absolutely dwarf anything that Congress ever did. If the American people are ever going to learn what is really going on with our economy, then it is absolutely imperative that they get educated about the Federal Reserve.

So, I ask does this scare you, in fact all the central banks globally take actions like this in not making sure free markets are truly free. In fact all people were taught that if you print more money your money is worth less, and that's what all central banks do globally, devalue our money supply so the poor remain poorer.

Source:https://www.facebook.com/photo.php?fbid=511686842177420&set=a.361837760495663.95686.176963112316463&type=1&theater

What causes the rise and fall of nations?

"This too shall pass"

I like what Juan Enriqez said in a TED talk:

"Your fear should be that you are not, that you're paying attention to stuff which is temporal. I mean, George Bush -- he's going to be gone, all right? Life isn't. Whether the humans survive or don't survive, these things are going to be living on this planet or other planets. And it's just beginning to understand this code of DNA that's really the most exciting intellectual adventure that we've ever been on."

Well I would like to add to that there's a lot more to that, I mean this election cycle, the next and all the fallacy of government bickering will pass and one politician will come to power and blame another. However, what they don't understand and what governments don't understand is the fundamentals that make an economy grow.

Juan on the next page goes on to mention:

The greatest repository of knowledge when most of us went to college

was this thing, and it turns out that this is not so important any more. The U.S. Library of Congress, in terms of its printed volume of data, contains less data than is coming out of a good genomics company every month on a compound basis. Let me say that again: A single genomics company generates more data in a month, on a compound basis, than is in the printed collections of the Library of Congress. This is what's been powering the U.S. economy. It's Moore's Law. So, all of you know that the price of computers halves every 18 months and the power doubles, right? Except that when you lay that side by side with the speed with which gene data's being deposited in GenBank, Moore's Law is right here: it's the blue line. This is on a log scale, and that's what superexponential growth means. This is going to push computers to have to grow faster than they've been growing because so far, there haven't been applications that have been required that need to go faster than Moore's Law. This stuff does.

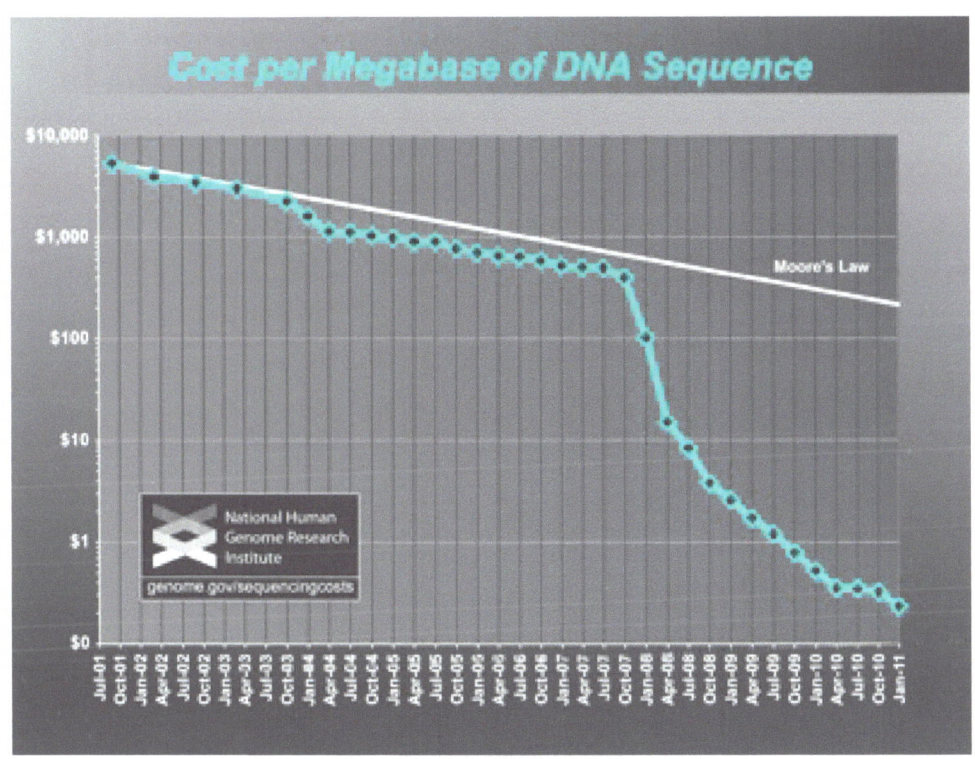

Juan goes on to mention in his TED talk:

And here's an interesting map. This is a map which was finished at the Harvard Business School. One of the really interesting questions is, if all this data's free, who's using it? This is the greatest public library in the world. Well, it turns out that there's about 27 trillion bits moving inside from the United States to the United States; about 4.6 trillion is going over to those European countries; about 5.5's going to Japan; there's almost no communication between Japan, and nobody else is literate in this stuff. It's free. No-one's reading it. They're focusing on the war; they're focusing on Bush; they're not interested in life. So, this is what a new map of the world looks like. That is the genomically literate world. And that is a problem.

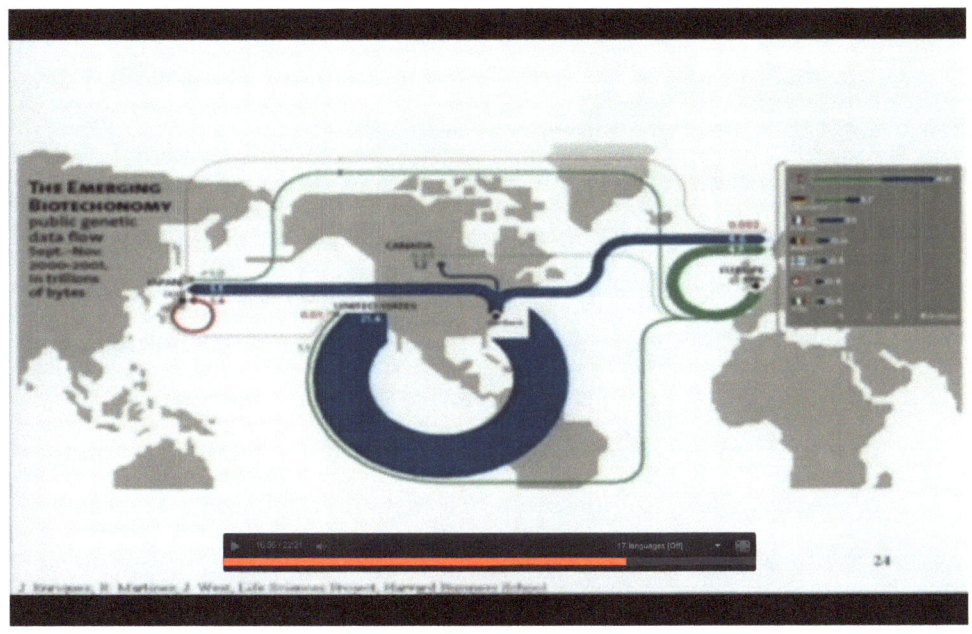

Source:

http://www.ted.com/talks/juan_enriquez_on_genomics_and_our_future.html

Graphs like this clearly explain the rise and fall of nations. I mean look at the technologically literate nations on the map, Europe, North America and Japan, are the only 3 major industrialized nations with overall equal access to technology and quality of life that's similar. This map shows the three nations where most of open genetic information is being used, and studied over the internet. The map above shows the genetically literate world. Of course Dubai, Hong Kong, Singapore, Monaco, Luxembourg, and a few tiny nation-states that have done well and as a result are very wealthy.

What are the implications of maps like this? Well it tells us very few nations are making use of the free data available and leveraging it for their benefit. For instance today to sequence a genome using an Ion Torrent machine it costs only 1000 dollars, and that's a full human sequence of the genome? How is that relevant you say? Well for that small price if you can sequence and study genes you can solve trillions of dollars' worth of answers and have genomic data worth millions and billions of dollar using very little people working in healthcare and information technology related fields. Maybe Juan says it best about what this means, when he laments, "That means you don't need to be a big nation to be successful; it means you don't need a lot of people to be successful; and it means you can move most of the wealth of a country in about three or four carefully picked 747s."

You wanna know what's immensely scary? I thought it was when Juan maintained in his TED talk that:

In an agricultural society, the difference between the richest and the poorest, the most productive and the least productive, was five to one. Why? Because in agriculture, if you had 10 kids and you grow up a little bit earlier and you work a little bit harder, you could produce about five times more wealth, on average, than your neighbor. In a knowledge society, that number is now 427 to 1. It really matters if you're literate, not just in reading and writing in English and French and German, but in Microsoft and Linux and Apple. And very soon it's going to matter if you're literate in life code. So, if there is something you should fear, it's that you're not keeping your eye on the ball. Because it really matters who speaks life. That's why nations rise and fall.

In fact this is not surprising as this chart reveals nations with high speed broadband and higher connectivity rates, especially with wireless broadband connections seem to have higher GDP per capita as this graph shows below these are the most connected countries and ironically the most prosperous nations per capita.

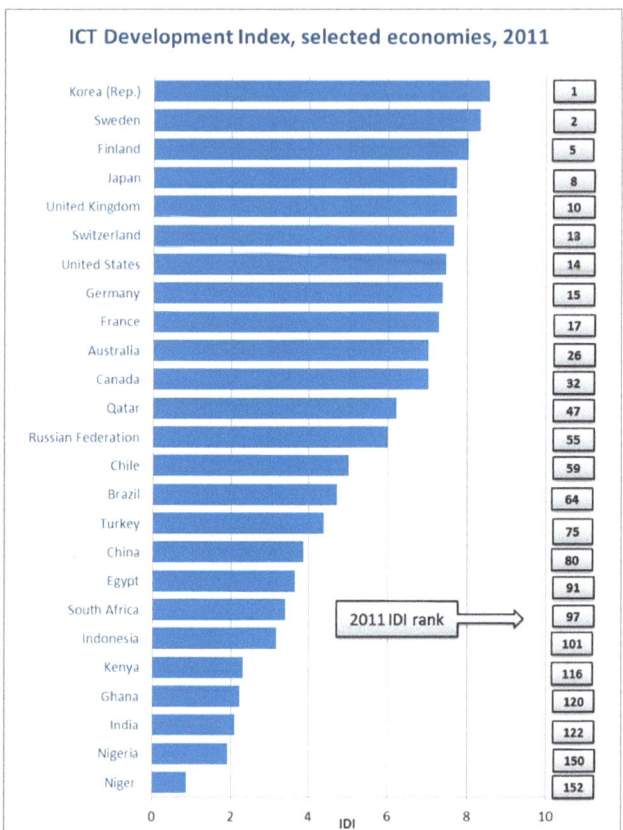

I've seen almost all of Juan's TED talks and talks at MIT. Every time I watch him speak I'm astonished at what I learn. However the quote from the talk above and the cause of nations to rise and fall kind of stuck out with me? I remember thinking that how do we get those nations that are poor or nations that are collapsing to rise? Well, know I know it has nothing to do with governments, politicians, credentials of public servants. You could take a Ph.D. Nobel Laureate economist that attended Harvard, Stanford and put them as the President of India, China or America and still the respective nations would collapse. Why? Different rules apply in the 21st century. The rules of the Keysian economists or mercantilism don't apply in today's world. In fact I like to say today anyone with a smartphone or tablet with access to all of human knowledge is actually living in the land as a kings. Each man, woman or child their own, with the power to create knowledge and participate in the global knowledge forum and communicate using video, text and email to anyone in the world. In fact with all this knowledge as my friend Dr. Peter Diamandis, founder of the X prize foundation says, a Massai warrior in Africa has access to more information than the President Bill Clinton did just 15 years ago!

So what are the conditions that make a developed nation? What would it look like? Maybe like South Korea, Taiwan, Singapore, or Hong Kong. These once impoverished nations and even at one time agriculturally dominated societies once had per capita incomes comparable to sub Saharan African countries, but today dominate the world.

One such example would be South Korea. According to Ben Goertzel, an Artificial intelligence researcher, in 1960, South Korea was poorer than two-thirds of the nations in sub-Saharan Africa. Today it's the world's most digital nation, with a per capita income of nearly $28,000, higher than New Zealand ($27K) or Portugal ($22K). This transformation largely took place during 1965-1985, and is known as the "Korean Miracle."

According to Ben, the reasons for Korea's success are quite simple, and quite in tune with what Juan Enriquez said about the fact that you don't need to be a big nation to be successful and you can move the wealth of entire nations on a few carefully picked Jet Aircraft. Ben goes on to mention that the factors underlying the first Korean Miracle are fairly well understood by social scientists. They include:

- a relatively equal distribution of wealth (which arose by historical accident, due to the Japanese occupation and the Korean war)
- a highly educated population
- a relatively young population (presently, e.g. 10% of Koreans and 20% of Germans are 65 or older)
- a culture oriented toward hard work
- a tradition of bureaucratic service without extreme levels of corruption
- high savings and investment rates
- a concerted effort on the part of the government to stimulate and protect a set of carefully selected manufacturing and technology industries

While these were great for a while, what's astonishing is that the next level of focus in South Korea is on:

- **Nano Materials and Devices Institute**
- **Integrated Bioscience and Biotechnology Institute**
- **Advanced Automotive Institute**
- **Intelligent Robotics Institute**
- **Software and Digital Media Institute**
- **Environment, Energy, Resources Institute**
- **Information Technology Institute**
- **Urban Infra-Tech Institute**
- **Transdisciplinary Studies Institute**

What's astonishing is not what the government has done, but rather what the government has not done, there is a sort of no government intervention and a sense of libertarianism in Korea, which to me means there isn't much persecution for studying modern science. As Ben Goertzel goes on to mention, "not only does the government provide systematic, well-conceived stimuli for advancing technologies, it also provides a conducive regulatory environment, unimpaired by the

Luddite "ethical" worries that plague research in many areas, particularly biomedicine, in the Western world. Cloning research has flourished, including the first successful cloning of a dog, Snuppy, and the cloning of two females of an endangered species of wolves at SNU. Stem cell research has been heavily funded, and I personally know a group of Americans currently working on setting up a stem cell therapy center in South Korea, aimed at life extension and the cure and prevention of aging-related diseases.

And the focus on advanced technology extends from the upper reaches of the educational system right down to the youngest children. Korea is the first country of the world to bring high-speed fiber-optic broadband to every primary and secondary school; and promises universal distribution of free digital textbooks by 2013."

In fact there number one company which accounts for a large share of South Korea's revenue, Samsung may be the first to have a domestic robot in every home by 2020! In fact they wanted a robot in most kindergartens by 2013! I have no doubt this freedom to pursue technology to do work and interact with machines will make the future very prosperous. In fact I'm confident we will have humanoids aid the elderly very soon and replace many factory workers, moving them up the value chain to programming or retraining or even learning new skills to make intellectual property, which will soon be the currency of the 21st century.

In fact rich countries that haven't taken their eye of the ball tend to prosper. In fact, Ben even goes on to mention that "there are equally fascinating possibilities on the nano-bio side. Samsung engineers recently filed a patent for carbon nanotube electron emitters, perhaps intended for use in a new generation of display screens. On the other hand, researchers in Israel and the US are now exploring the viability of carbon nanotubes as a tool for brain-computer interfacing: connecting computer chips to brain matter. The convergence possibilities here are obvious. Samsung cranial implants, coming soon to an Interweb near you? Or more prosaically, how about bold new devices allowing the blind to see, the deaf to hear far more clearly than cochlear implants, the financial trader and the biologist to experience the markets and the genome as immediately as we perceive the trees in the forest."

Imagine the breakthroughs and the trillions in wealth and lives they'll save while other countries burn to dust!

Dr. Peter Diamandis, founder of the X Prize foundation and Singularlty University says it best when he maintains on a blog post that people and corporations that don't keep their eyes on the ball of exponential technologies will fail.

In fact in a recent blog post he maintained:
"The year 2012 marks the death of Kodak, a $26-billion, century-old "cornerstone" company of the U.S. R.I.P. Did you know that Kodak actually invented the digital camera that ultimately put it out of business? Kodak had the patents and a head start, but ignored all that. Why? That's what this blog is about.

To put an exclamation mark at the end of the Kodak story: In this same year, Instagram, another company in the image business, was acquired by Facebook for $1 billion… The catch is, Instagram had only 13 employees at the time. This is the difference between a "linear" company and an "exponential" one.

One of the biggest challenges that large companies have today is creating an environment that allows for innovation. Everywhere the rate of change is so fast that large U.S. companies are in constant danger of disruption. Not from competition in China or India, no.

They're in danger of being made obsolete from two guys/gals in a garage in Silicon Valley, or anyone, anywhere, empowered by exponential technology, willing to risk it all, driven by their passion."

This may sound scary to an outside businessman because this is what's happening every day. Great companies are failing only to be displaced by kids in garages that are being funded for outlandish ideas and taking on the incumbents. Whether that's Square replacing old credit card readers, Spotify replacing the big old record labels, or Rovio, the company that makes Angry Birds taking on old school gaming companies, or YouTube fighting the outdated movie and television studios.

So, the real bugger now! You know what happens to countries that take their eyes off the ball? Well they become like India today. $1/5^{th}$ of the population moving less than 5% of the world's wealth. In the mid part of this year in 2012 in July's power outage in India, it affected 670 million people, a whopping 10 percent of the global population. Trains stopped, coal miners were stranded underground, people were left without water, and even crematoriums faced the uncomfortable problem of halting their business.

The technology website GigaOM maintained, three of the country's interconnected northern power grids collapsed, covering an area that extended 2,000 miles. Politicians blamed northern states for consuming too much power, essentially overdrawing their allocation from the national grid. More than half of India's power comes from coal, and it is widely reported that populist policies force. India's utilities to sell power below the cost of generating it, which creates a system where the public utilities are in major debt. That ultimately means the grid never gets

upgraded.

Furthermore, according to GigaOM, India had been hoping to pay about $40 per ton for coal, but it is now finding that paying $120 per ton for imported coal is about as good a deal as it's going to see. Indian power companies have scrapped 42 gigawatts of new power plants, partially due to warnings from the Reserve Bank of India that coal projects carried higher risks. India actually has coal deposits, but transporting it is difficult, since most of its deposits are distant from India's booming coastal cities.

The question, of course is, what now? India's growing middle class and faster growing

economy are demanding power generation on a new scale. One of the leading indicators of the burgeoning middle class there is the 20 percent annual growth rate for air-conditioning sales. In fact its problems like these that happen, while Singapore, and South Korea and Japan have freer markets for energy, India has nationalized ancient coal. In fact there's power cuts and water "shortages" that are superficial. In fact using desalination plants, they could give over a billion Indians plenty of water, but foreign investment is far and few. India is a real bugger, because power is unstable, the grid is broken and still a large number of people live off the grid. So not having a true free market has ruined the nation. The totalitarian socialism and governments ruin countries. Well maybe that's what happens you when you don't pay attention to growth and allow free market technologies to solve problems. Anyways, my third chapter outlines how the future will have an abundance of power very soon and how all our energy is getting cheaper and growing exponentially soon. While all these "shortages" will be solved soon at hand.

Since we're entitled and all have the power to build great empires, why not just break away from these fiat currencies and stupid laws and regulations and taxes and wars and start up a country like we start up a country, just like we start up great companies to solve problems. Well that's exactly the motive behind seasteding! Sound cool yet? Pay attention.

Source: http://www.economist.com/node/21540395

So here's a model of what a possible floating city would look like and yes since we have cruise ships and floating oil platforms they would be self-sustained and self-powered with wind turbines or other sources of power like natural gas generators and desalinate water and solve the little problems and weather storms. Since cruise ships and oil platforms are safe and non-interventionist they'd do well against attack and since people would be armed on these cities they could theoretically defend themselves well. According to the Economist magazine, they believe that in the same way the Pilgrims who set out from England on the *Mayflower* to escape an intolerant, over-mighty government and build a new society were lucky to find plenty of land in the New World on which to build it, they truly believe that some modern libertarians, such as Peter Thiel, one of the founders of PayPal, dream of setting sail once more to found colonies of like-minded souls. They claim, however, all the land on Earth has been claimed by the governments they seek to escape. So, they conclude, they must build new cities on the high seas, known as seasteads. There are many fans of this idea, myself included but also Patri Friedman, yes the grandson of Milton Friedman and we believe it's possible and happening. In fact just recently Thiel gave 1.25 million towards the seasteding institute. Since I've told you the conditions required to build a knowledge based society that's rich and technology based, free of all laws and free of bureaucracy, the smart people may flee nations and start companies on seasteding colonies. In fact these may even serve as experimental hotbeds of different democracies and governments and see which models work well and fail. Since we know the richest nation on Earth in tiny landlocked Luxembourg with a 115000 dollar per capita income in a tiny landlocked nation with no oil, coal or diamonds or any valuable resources moving 100x wealth per capita compared to India and more than twice than an average American, we know that just smart people on the internet can build nations of technological marvel and abundance that are research hotbeds. I believe smart people in the 21st century will pursue this to further genomics, human cloning, mind uploading and even space travel with offshore launch platforms and research labs that will make tons of money on the internet and sell products and goods and buy resources from nations that they lack. These will be true free societies devoid of any rules.

Seastead advocates are not crazed anarchists against government. They are promoting new forms of governance.

For instance, in his paper "Governing Seasteads: An Outline of the Options," Brad Taylor, A well noted advocate of seasteding observed: "We tend to take the existence of states—monopolistic providers of governance and wielders of coercive force over some large geographic area—for granted. Throughout most of human history, though, rules have been created and enforced in a decentralized way, producing customary systems of law." Taylor reviews many of the options which exist outside of formal civil government.

In fact one such concept will happen very soon called Blueseed. An amazing combination of science fiction and governmental failure has brought about a great opportunity for an entrepreneurial incubator, just 12 miles off the California coastline! The Founders of Blueseed have raised a large amount of startup capital from venture capitalists including Peter Thiel (founder of Paypal) to purchase a decommissioned cruise ship, retrofit it to support the life and work of 1000+ entrepreneurs and highly intelligent foreign nationals from around the globe, and create a small village at sea for the purpose of building the next Silicon Valley super-startup free of restrictive US Visa laws and regulations. This is an amazing use of innovative thinking to circumvent the inefficiencies of government! The boldest, brightest, and most talented tech entrepreneurs from around the world. Plus the individuals and organizations that support and invest in them. In fact over 950+ people have already expressed interest. What's the cost? They say living expenses are just 1500 a month, a bargain compared to San Francisco. A lot of people have visa problems or limitations to start a company and have to pay high taxes, by being an offshore entity, anyone from anywhere can work on these libertarian Islands and then even later move back to their homeland or transition to America and get help to get a Visa or so forth. Often times I meet entrepreneurs who move here who can't legally start a company or work more than a certain time, and this hurts all of us economically, this solution is a great idea to circumvent this.

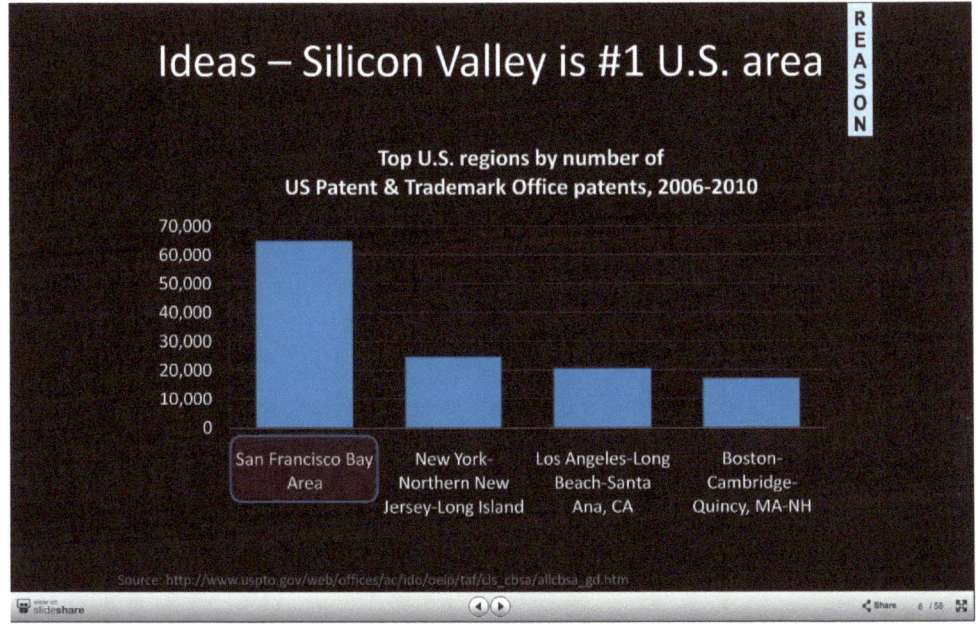

So maybe the reason why overall America is still the richest country and people still flock to Silicon Valley. However, what if you could get the benefits of Silicon Valley without the drawbacks? Well that's probably why it's located so close to the Valley. So it offers a win, win and win solution to citizens. So if the Silicon valleys of the world is where the biggest breakthroughs happen, then what's in the pipeline? Well the next chapter will blow you out of your mind on what's coming to those who keep their eye on the ball!

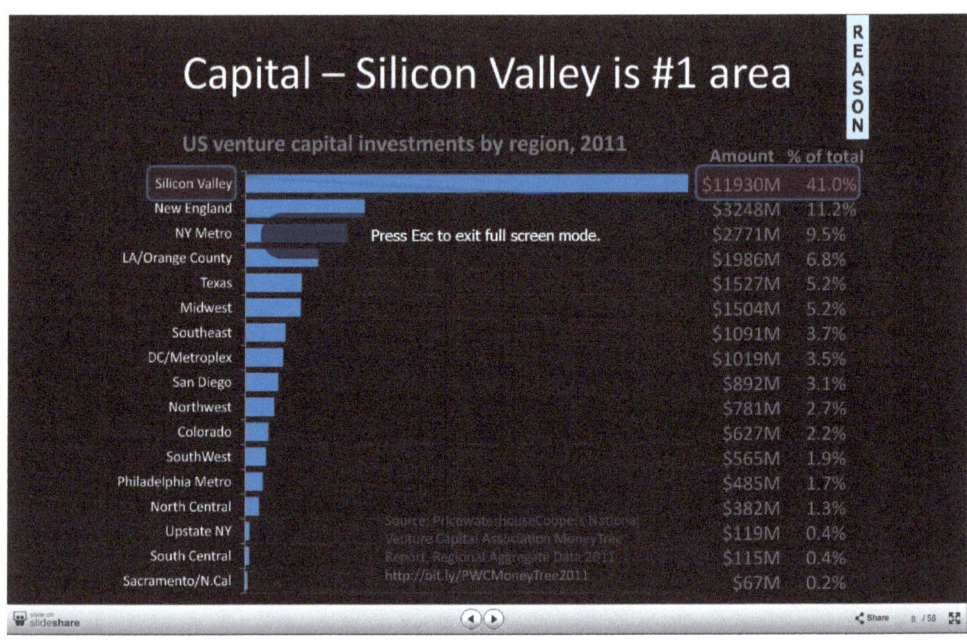

SAGA 2

TECHNOLOGY EVOLUTION

"We'll have billions of nanobots in the capillaries of our brains, communicating wirelessly with our biological neurons, with the Internet, with each other, and basically expanding human intelligence and experience. It's not some alien invasion of intelligent machines. It's coming from within our civilization. It expands our own intellectual powers." — Ray Kurzweil

An article I once read once said that, futurology is somewhat an acquired taste. Either, you're the kind of person who enjoys thinking about hotels on the moon, taking elevators to space and robots with human brains, or you are not. Either you're an optimist that thinks autonomous cars are cool and extremely convenient when you don't want to drive every day or you think Self-driving vehicles may put millions of truck drivers out of business and think that news articles and reports generated by algorithms take human writers out of the equation and it's not so good. You think geez, Language translation is now a free, online service and its great or you think wow lots of people will be out of jobs.

When predicting and talking about the future these are the two types of people you meet. Either they are the technology optimists who believe it's inevitable and we are already merging with our machines or they are deeply afraid of change. I truly believe you should be excited about the evolution of technology. First of all I'll explain the ideas of Ray Kurzweil, the prominent inventor, futurist, and author of books like "The Singularity is Near" and "How to Create a Mind The Secret of Human thought Revealed."

According to Ray Kurzweil the history of technology shows that technological change is exponential, contrary to the common-sense "intuitive linear" view. So according to his prediction we won't experience 100 years of progress in the 21st century, instead it will be 20,000 years of progress. The "returns," such as computer chip speed and cost-effectiveness, also increase exponentially. In fact Kurzweil says that there's even exponential growth in the rate of exponential growth. Within a few decades, machine intelligence will surpass human intelligence, leading to The Singularity — technological change so rapid and profound it represents a rupture in the fabric of human history. The implications include the merger of biological and non-biological intelligence, immortal software-based humans, and ultra-high levels of intelligence that expand outward in the universe at the speed of light.

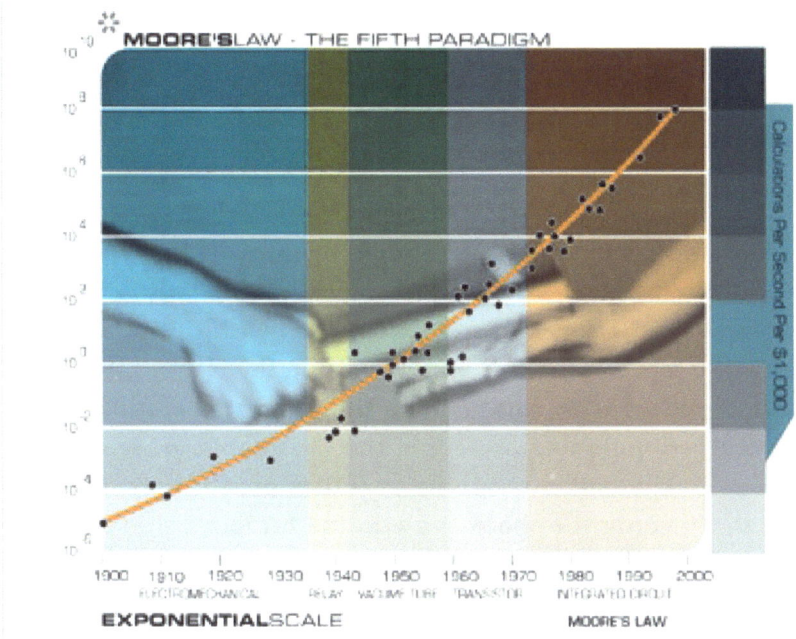

Source: http://www.kurzweilai.net/the-law-of-accelerating-returns

     According to Ray Kurzweil as the graph above illustrates Moore's law was only the fifth paradigm of exponential computation. In fact Moore's law may end soon, but that won't be the end of exponential growth of computation as we will switch to 3D circuits as we live in the third dimension. In fact Kurzweil believes that "law of accelerating returns" applies to all of technology, indeed to any true evolutionary process, and can be measured with remarkable precision in information based technologies. He cites examples of this exponential growth implied by the law of accelerating returns in technologies as varied as DNA sequencing, communication speeds, electronics of all kinds, and even in the rapidly shrinking size of technology. The Singularity results not from the exponential explosion of computation alone, but rather from the interplay and myriad synergies that will result from manifold intertwined technological revolutions. Also, keep in mind that every point on the exponential growth curves underlying these panoply of technologies (see the graphs below) represents an intense human drama of innovation and competition. It is remarkable therefore that these chaotic processes result in such smooth and predictable exponential trends.

Popular Mechanics recently did a great article entitled "110 Predictions for the Next 110 years," and from what it looks it appears to be a conservative prediction that resonates well with that of famed futurist Ray Kurzweil and Michio Kaku as well as a few others I've been following and deeply with what I think the projected trends of certain exponential technologies. Below is the time table for what Popular mechanics believes is a roadmap of future predictions. I've elaborated later in this chapter about the various technologies that's already making these technologies once stuck in the realm of science fiction into science fact.

WITHIN 20 YEARS...
<u>Self-driving cars</u> will hit the mainstream market.
Battles will be waged without direct human participation (think robots or unmanned aerial vehicles).
The first fully functional brain-controlled bionic limb will arrive.

WITHIN 30 YEARS...
All-purpose robots <u>will help us with household chores</u>.
Space travel will become as affordable as a round-the-world plane ticket.
<u>Soldiers will use exoskeletons</u> to enhance battlefield performance.

WITHIN 40 YEARS...
Nanobots will perform medical procedures inside our bodies.

WITHIN 50 YEARS...
We will have a colony on Mars.
Doctors will successfully transplant a lab-grown human heart.
We will fly the friendly skies without pilots onboard.

And renewable energy sources will surpass fossil fuels in electricity generation.

WITHIN 60 YEARS...
Digital data (texts, songs, etc.) will be zapped directly into our brains.
We will activate the first fusion power plant.
And we will wage the first battle in space.

Singularity

"I set the date for the Singularity -representing a profound and disruptive transformation in human capability- as 2045. The nonbiological intelligence created in that year will be one billion times more powerful than all human intelligence today." — Ray Kurzweil

I think Ian Barry, a staff writer for the student newspaper, *The Elm* says it best when he says Medical science has come a long way in the past 20 years. Diseases that were once thought unassailable have become treatable and manageable. Gene therapy has recovered from its setbacks and is showing promise in curing a number of diseases. Prosthetics have developed from what was essentially a shoe on a stick to fully articulated robotic limbs, controlled through neural interfaces and capable of sensory feedback. The price: performance ratio of gene sequencing is falling at an exponential rate. More and more sophisticated neural implants are being developed, allowing the human brain to interface directly with a computer. In fact I've personally worked in a brain machine interface lab where, even using a non-invasive Brain Machine interfaces, especially the P300 a system that detects oddball changes in the brain or "rare" events, we were able to type utilizing thought alone or even update my twitter status. In fact other friends of mine were able to control a wheelchair totally using thought alone using a 150 dollar headset from Emotive, the EPOC and even control a robotic arm to pick up and drink a soda using thought alone. In fact one student at another institution even found a way to fly a remote control quadricopter using thought alone as pictured below.

So what is the singularity? In physics, it represents the unknown point of a black hole where things get weird shall we say for argument sake.

In technological terms however, it's a term coined by Vernor Vinge and popularized by famed inventor and futurist Ray Kurzweil.

It can be summed as three basic principles all leading up to the idea of a greater than Human INTELLIGENCE!

1. *Intelligence explosion*: When humanity builds machines with greater-than-human intelligence, they will also be better than we are at creating still smarter machines. Those improved machines will be even *more* capable of improving themselves or their successors. This is a positive feedback loop that could, before losing steam, produce a machine with vastly greater than human intelligence: *machine superintelligence*. Such a superintelligence would have enormous powers to make the future unlike anything that came before it.
2. *Event horizon*: All social and technological progress thus far has come from human brains. When technology creates entirely new kinds of intelligence, this will cause the future to be stranger than

we can imagine. So there is an 'event horizon' in the future beyond which our ability to predict the future rapidly breaks down.
3. *Accelerating change*: Technological progress is faster today than it was a century ago, and it was faster a century ago than it was 500 years ago. Technological progress feeds on itself, leading to accelerating change much faster than the linear change we commonly expect, and perhaps change that is faster than we can cope with.

According to Michael Anissimov, what are the benefits of the successful singularity? Anissimov states that If human researchers end up successfully creating general intelligence in a machine, the event would constitute our first encounter with a true "alien intelligence" that is our equal. This article will assume that the programmers of the first artificial intelligences have successfully solved the extremely difficult problems of both describing intelligence in terms of algorithms and ensuring that the resulting minds are explicitly human-friendly. If they succeeded, what would be the benefits?

According to Anissimov, One of the greatest limitations of human intelligence is its speed. Our neurons fire about 200 times per second under ideal conditions. In contrast, silicon transistors can send and receive electronic signals more than 2 billion times per second. Our brains are massively parallel, with about a hundred billion neurons, but each neuron operates so slowly that a $10,000 desktop supercomputer can execute 933 billion operations per second, which is within a factor of 100,000 of one of the most common estimates of the processing power of the human brain, 10^{17} operations per second. Several supercomputers have already passed the 10^{15} operations per second mark.

I do agree with Michael when he maintains that most artificial intelligence researchers agree that the primary challenge is not hardware, but software and that if we understood the processes of intelligence in enough detail to implement them on computers, it seems likely that ample hardware would be available to create many copies. These copies could also be run in an accelerated fashion on distributed systems. This "instant intelligence, just add computing power" principle is a recipe for explosive economic growth.

According to Michael he states, "Artificial intelligences with human interests in mind could use their machine minds to solve human challenges in fields like medicine, physics, chemistry, engineering, politics, diplomacy, biology, sociology, and economics. Being native to the world of computers, they could run complex simulations in mere moments that would take human researchers years to build. Complex, detailed, mathematically accurate simulations could be the default thought mode for artificial intelligences-their "thoughts" could be far superior to our best simulations."

Even more he maintains that having direct access to all the freely available data and information on the internet, combined with complex internal simulations, and eventually accompanied by direct experiments through robotic proxies, the first roughly human-equivalent artificial intelligences could contribute huge value to humanity within a very short time.

Thus, what does this mean?

It means that since machines will soon surpass us in intelligence we will have no choice but to merge with the machines. Raymond Kurzweil, the prominent inventor, author and futurist in fact writes about building the mind so we can start replacing our own sections of our brain with non-biological substrates so we can become part machine. Sound scary? Not quite.

The primary thesis of his book is that our brains higher level of intelligence comes from the Neo-cortex. In fact I've taken a class on neurobiology at Harvard and can confirm this idea that the neocortex is so folded and takes up much of the area of the brain which gives rise to higher levels of human intelligence. One more thing, the neocortex is actually based on massive parallel redundancy. Just like many repetitive transistors in supercomputers working together. Ultimately, Kurzweil believes that we will build an artificial neocortex that has the full range of capability as the human counterpart. In fact according to Kurzweil electronic circuits are millions of times faster than biological counterparts. He says as we replace our biological neocortex with a digital one, we won't have to worry about how much memory we can store or loosing information that's stored in our circuits. In fact we both predict that most of our computation and non-daily use information will be shifted to the cloud to which can we can have instant recall or long term storage of memories we can access. He estimates we have 300 million pattern recognizers stored in our brain in the neocortex, if we augment with nano-technology based replacements we can store an infinitely large amount or as much computation as the universe allows to store information as we please. However, one big advantage of replace biological neurons with synthetic digital ones with identical function is the information can be backed up in real time to the cloud so that we can infinitely replicate our information, and our personalities we acquire over time. In fact we can clone multiple avatars of ourselves in the cloud and live digitally in virtually reality that will be a simulated version of real reality. In fact this is the goal of the singularity and mind uploading. I believe building a digital neocortex is the first step in this endeavor to replacing our neurons to digital from which we can digitally replicate ourselves forever.

Super-human artificial intelligence is the creation of greater-than-human level knowledge, reasoning, and cognition in a computer. The practical applications of SAI in human activities such as science, engineering, politics, business, medicine, and entertainment are almost without limit.

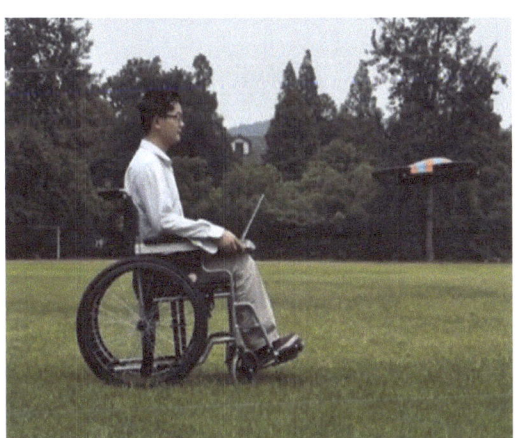

Source: http://www.thinktechuk.com/blog/wp-content/uploads/2012/09/EmotivAR-Dronewheel-chair.jpg

A lot of futurists get this wrong, when talking about brain machine interfaces because they are barely aware of the numerous technologies and advancements taking place daily. They always think highly advanced mind control is a far off thing but in fact the technology is growing exponentially. I could point to over 100 different applications and technologies of BCI personally and uses for people that would dazzle you, in fact I'll highlight a few that I personally love.

One of them would be using a BCI interface to help drive a car. One of the applications is called brain driver and its quite useful when used in combination with an autonomous car. In fact that's how they use it. The researchers caution that the BrainDriver application is still a demonstration and is not ready for the road. But they say that future human-machine interfaces like this have huge potential to improve driving, especially in combination with autonomous vehicles. As an example, they mention an autonomous cab ride, where the passenger could decide, only by thinking, which route to take when more than one possibility exist. Or another scenario would be upon arrival

Source: http://spectrum.ieee.org/automaton/robotics/robotics-software/braindriver-a-mind-controlled-car

This image shows a driver wearing a 200 dollar Emotiv Epoc Headset and controlling an autonomous car in mind control mode. The researchers caution that the BrainDriver application is still a demonstration and is not ready for the road. But they say that future human-machine interfaces like this have huge potential to improve driving, especially in combination with autonomous vehicles. As an example, they mention an autonomous cab ride, where the passenger could decide, only by thinking, which route to take when more than one possibility exist.

This type of non-invasive brain interface could also allow disabled and paralyzed people to gain more mobility in the future, similarly to what is already happening in applications such as robotic exoskeletons and advanced prosthetics.

Thus, Raúl Rojas, an AI professor at the Freie Universität Berlin, and his team have demonstrated how a driver can use a brain machine interface to steer a vehicle. Here's what the researchers say about the project, which they call the BrainDriver:

"After testing iPhone, iPad and an eye-tracking device as possible user interfaces to maneuver our research car, named "MadeInGermany," we now also use Brain Power. The "BrainDriver" application is of course a demonstration and not roadworthy yet, but in the long run human-machine interfaces like this could bear huge potential in combination with autonomous driving."

Maybe these three laws show why most "scientists" are wrong.

Clarke's Three Laws are three "laws" of prediction formulated by the British writer Arthur C. Clarke. They are:

1. When a distinguished but elderly scientist states that something is possible, he is almost certainly right. When he states that something is impossible, he is very probably wrong.

2. The only way of discovering the limits of the possible is to venture a little way past them into the impossible.

3. Any sufficiently advanced technology is indistinguishable from magic.

SINGULARITY

Our next saga in human evolution is the singularity itself. What is the singularity and what is a future scenario going to look like? Well maybe of all the people on earth, one of the surprising mainstream proponents in Glenn Beck! Yep that one, the one people love to hate. Let me put an excerpt from a final chapter of his book summarizes the why you shouldn't fear the singularity.

Here's an excerpt from the final chapter of his new book Cowards, titled "*Adapt or Die: The Coming Intelligence Explosion:*"

> *The year is 1678 and you've just arrived in England via a time machine. You take out your new iPhone in front of a group of scientists who have gathered to marvel at your arrival.*
>
> *"Siri," you say, addressing the phone's voice-activated artificial intelligence system, "play me some Beethoven."*

Dunh-Dunh-Dunh-Duuunnnhhh! The famous opening notes of Beethoven's Fifth Symphony, stored in your music library, play loudly.

"Siri, call my mother."

Your mother's face appears on the screen, a Hawaiian beach behind her. "Hi, Mom!" you say. "How many fingers am I holding up?"

"Three," she correctly answers. "Why haven't you called more—"

"Thanks, Mom! Gotta run!" you interrupt, hanging up.

"Now," you say. "Watch this."

Your new friends look at the iPhone expectantly.

"Siri, I need to hide a body."

Without hesitation, Siri asks: "What kind of place are you looking for? Mines, reservoirs, metal foundries, dumps, or swamps?" (I'm not kidding. If you have an iPhone 4S, try it.)

You respond "Swamps," and Siri pulls up a satellite map showing you nearby swamps.

The scientists are shocked into silence. What is this thing that plays music, instantly teleports video of someone across the globe, helps you get away with murder, and is small enough to fit into a pocket?

At best, your seventeenth-century friends would worship you as a messenger of God. At worst, you'd be burned at the stake for witchcraft. After all, as science fiction author Arthur C. Clarke once said, "Any sufficiently advanced technology is indistinguishable from magic."

Now, imagine telling this group that capitalism and representative democracy will take the world by storm, lifting hundreds of millions of people out of poverty. Imagine telling them their descendants will eradicate smallpox and regularly live seventy-five or more years. Imagine telling them that men will walk on the moon, that planes, flying hundreds of miles an hour, will transport people around the world, or that cities will be filled with buildings reaching thousands of feet into the air.

They'd probably escort you to the madhouse.

Unless, that is, one of the people in that group had been a man named Ray Kurzweil.

Kurzweil is an inventor and futurist who has done a better job than most at predicting the future. Dozens of the predictions from his 1990 book *The Age of Intelligent Machines* came true during the 1990s and 2000s. His follow-up book, *The Age of Spiritual Machines*, published in 1999, fared even better. Of the 147 predictions that Kurzweil made for 2009, 78 percent turned out to be entirely correct, and another 8 percent were roughly correct. For example, even though every portable computer had a keyboard in 1999, Kurzweil predicted that most portable computers would lack a keyboard by 2009. It turns out he was right: by 2009, most portable computers were MP3 players, smartphones, tablets, portable game machines, and other devices that lacked keyboards.

Kurzweil is most famous for his "law of accelerating returns," the idea that technological progress is generally "exponential" (like a hockey stick, curving up sharply) rather than "linear" (like a straight line, rising slowly). In nongeek-speak that means that our knowledge is like the compound interest you get on your bank account: it increases exponentially as time goes on because it keeps building on itself. We won't experience one hundred years of progress in the twenty-first century, but rather twenty thousand years of progress (measured at today's rate).

Many experts have criticized Kurzweil's forecasting methods, but a careful and extensive review of technological trends by researchers at the Santa Fe Institute came to the same basic conclusion: technological progress generally tends to be exponential (or even faster than exponential), not linear.

So, what does this mean? In his 2005 book The Singularity Is Near, Kurzweil shares his predictions for the next few decades:

- In our current decade, Kurzweil expects real-time translation tools and automatic house-cleaning robots to become common.
- In the 2020s he expects to see the invention of tiny robots that can be injected into our bodies to intelligently find and repair damage and cure infections.
- By the 2030s he expects "mind uploading" to be possible, meaning that your memories and personality and consciousness could be copied to a machine. You could then make backup copies of yourself, and achieve a kind of technological immortality.

Age of the Machines?

"We became the dominant species on this planet by being the most intelligent species around. This century we are going to cede that crown to machines. After we do that, it will be them steering history rather than us."

—Jaan Tallinn, co-creator of Skype and Kazaa

If any of that sounds absurd, remember again how absurd the eradication of smallpox or the iPhone 4S would have seemed to those seventeenth-century scientists. That's because the human brain is conditioned to believe that the past is a great predictor of the future. While that might work fine in some areas, technology is not one of them. Just because it took decades to put two hundred transistors onto a computer chip doesn't mean that it will take decades to get to four hundred. In fact, Moore's Law, which states (roughly) that computing power doubles every two years, shows how technological progress must be thought of in terms of "hockey stick" progress, not "straight line" progress. Moore's Law has held for more than half a century already (we can currently fit 2.6 billion transistors onto a single chip) and there's little reason to expect that it won't continue to.

But the aspect of his book that has the most far-ranging ramifications for us is Kurzweil's prediction that we will achieve a "technological singularity" in 2045. He defines this term rather vaguely as "a future period during which the pace of technological change will be so rapid, its impact so deep, that human life will be irreversibly transformed."

Part of what Kurzweil is talking about is based on an older, more precise notion of "technological singularity" called an intelligence explosion. An intelligence explosion is what happens when we create artificial intelligence (AI) that is better than we are at the task of designing artificial intelligences. If the AI we create can improve its own intelligence without waiting for humans to make the next innovation, this will make it even more capable of improving its intelligence, which will . . . well, you get the point. The AI can, with enough improvements, make itself smarter than all of us mere humans put together.

The really exciting part (or the scary part, if your vision of the future is more like the movie The Terminator) is that, once the intelligence explosion happens, we'll get an AI that is as superior to us at science, politics, invention, and social skills as your computer's calculator is to you at arithmetic. The problems that have occupied mankind for decades— curing diseases, finding better energy sources, etc.— could, in many cases, be solved in a matter of weeks or months.

Again, this might sound far-fetched, but Ray Kurzweil isn't the only one who thinks an intelligence explosion could occur sometime this century. Justin Rattner, the chief technology officer at Intel, predicts some kind of Singularity by 2048. Michael Nielsen, co-author of the leading textbook on quantum computation, thinks there's a decent chance of an intelligence explosion by 2100. Richard Sutton, one of the biggest names in AI, predicts an intelligence explosion near the middle of the century. Leading philosopher David Chalmers is 50 percent confident an intelligence explosion will occur by 2100. Participants at a 2009 conference on AI tended to be 50 percent confident that an intelligence explosion would occur by 2045.

If we can properly prepare for the intelligence explosion and ensure that it goes well for humanity, it could be the best thing that has ever happened on this fragile planet. Consider the difference between humans and chimpanzees, which share 95 percent of their genetic code. A relatively small difference in intelligence gave humans the ability to invent farming, writing, science, democracy, capitalism, birth control, vaccines, space travel, and iPhones— all while chimpanzees kept flinging poo at each other.

Intelligent Design?

The thought that machines could one day have superhuman abilities should make us nervous. Once the machines are smarter and more capable than we are, we won't be able to negotiate with them any more than chimpanzees can negotiate with us. What if the machines don't want the same things we do?

The truth, unfortunately, is that every kind of AI we know how to build today definitely would not want the same things we do. To build an AI that does, we would need a more flexible "decision theory" for AI design and new techniques for making sense of human preferences. I know that sounds kind of nerdy, but AIs are made of math and so math is really important for choosing which results you get from building an AI.

These are the kinds of research problems being tackled by the Singularity Institute in America and the Future of Humanity Institute in Great Britain. Unfortunately, our silly species still spends more money each year on lipstick research than we do on figuring out how to make sure that the most important event of this century (maybe of all human history)— the intelligence explosion— actually goes well for us.

Likewise, self-improving machines could perform scientific experiments and build new technologies much faster and more intelligently than humans can. Curing cancer, finding clean energy, and extending life expectancies would be child's play for them. Imagine living out your own personal fantasy in a different virtual world every day. Imagine exploring the galaxy at near light speed, with a few backup copies of your mind safe at home on earth in case you run into an exploding supernova. Imagine a world where resources are harvested so efficiently that everyone's basic needs are taken care of, and political and economic incentives are so intelligently fine-tuned that "world peace" becomes, for the first time ever, more than a Super Bowl halftime show slogan.

With self-improving AI we may be able to eradicate suffering and death just as we once eradicated smallpox. It is not the limits of nature that prevent us from doing this, but only the limits of our current understanding. It may sound like a paradox, but it's our brains that prevent us from fully understanding our brains.

Turf Wars

At this point you might be asking yourself: "Why is this topic in this book? What does any of this have to do with the economy or national security or politics?"

In fact, it has everything to do with all of those issues, plus a whole lot more. The intelligence explosion will bring about change on a scale and scope not seen in the history of the world. If we don't prepare for it, things could get very bad, very fast. But if we do prepare for it, the intelligence explosion could be the best thing that has happened since . . . literally ever.

But before we get to the kind of life-altering progress that would come after the Singularity, we will first have to deal with a lot of smaller changes, many of which will throw entire industries and ways of life into turmoil. Take the music business, for example. It was not long ago that stores like Tower Records and Sam Goody were doing billions of dollars a year in compact disc sales; now people buy music from home via the Internet. Publishing is currently facing a similar upheaval. Newspapers and magazines have struggled to keep subscribers, booksellers like Borders have been forced into bankruptcy, and customers are forcing publishers to switch to ebooks faster than the publishers might like.

All of this is to say that some people are already witnessing the early stages of upheaval firsthand. But for everyone else, there is still a feeling that something is different this time; that all of those years of education and experience might be turned upside down in an instant. They might not be able to identify it exactly but they realize that the world they've known for forty, fifty, or sixty years is no longer the same.

There's a good reason for that. We feel it and sense it because it's true. It's happening. There's absolutely no question that the world in 2030 will be a very different place than the one we live in today. But there is a question, a large one, about whether that place will be better or worse.

It's human nature to resist change. We worry about our families, our careers, and our bank accounts. The executives in industries that are already experiencing cataclysmic shifts would much prefer to go back to the way things were ten years ago, when people still bought music, magazines, and books in stores. The future was predictable. Humans like that; it's part of our nature.

But predictability is no longer an option. The intelligence explosion, when it comes in earnest, is going to change everything— we can either be prepared for it and take advantage of it, or we can resist it and get run over.

Unfortunately, there are a good number of people who are going to resist it. Not only those in affected industries, but those who hold power at all levels. They see how technology is cutting out the middlemen, how people are becoming empowered, how bloggers can break national news and YouTube videos can create superstars.

And they don't like it.

A Battle for the Future

Power bases in business and politics that have been forged over decades, if not centuries, are being threatened with extinction, and they know it. So the owners of that power are trying to hold on. They think they can do that by dragging us backward. They think that, by growing the public's dependency on government, by taking away the entrepreneurial spirit and rewards and by limiting personal freedoms, they can slow down progress.

But they're wrong. The intelligence explosion is coming so long as science itself continues. Trying to put the genie back in the bottle by dragging us toward serfdom won't stop it and will, in fact, only leave the world with an economy and society that are completely unprepared for the amazing things that it could bring.

Robin Hanson, author of "The Economics of the Singularity" and an associate professor of economics at George Mason University, wrote that after the Singularity, "The world economy, which now doubles in 15 years or so, would soon double in somewhere from a week to a month."

That is unfathomable. But even if the rate were much slower, say a doubling of the world economy in two years, the shock-waves from that kind of growth would still change everything we've come to know and rely on. A machine could offer the ideal farming methods to double or triple crop production, but it can't force a farmer or an industry to implement them. A machine could find the cure for cancer, but it would be meaningless if the pharmaceutical industry or Food and Drug Administration refused to allow it. The machines won't be the problem; humans will be.

And that's why I wanted to write about this topic. We are at the forefront of something great, something that will make the Industrial Revolution look in comparison like a child discovering his hands. But we have to be prepared. We must be open to the changes that will come, because they will come. Only when we accept that will we be in a position to thrive. We can't allow politicians to blame progress for our problems. We can't allow entrenched bureaucrats and power-hungry executives to influence a future that they may have no place in.

Many people are afraid of these changes— of course they are: it's part of being human to fear the unknown— but we can't be so entrenched in the way the world works now that we are unable to handle change out of fear for what those changes might bring.

Change is going to be as much a part of our future as it has been of our past. Yes, it will happen faster and the changes themselves will be far more dramatic, but if we prepare for it, the change will mostly be positive. But that preparation is the key: we need to become more well-rounded as individuals so that we're able to constantly adapt to new ways of doing things. In the future, the way you do your job may change four to five or fifty times over the course of your life. Those who cannot, or will not, adapt will be left behind.

At the same time, the Singularity will give many more people the opportunity to be successful. Because things will change so rapidly there is a much greater likelihood that people will find something they excel at. But it could also mean that people's successes are much shorter-lived. The days of someone becoming a legend in any one business (think Clive Davis in music, Steven Spielberg in movies, or the Hearst family in publishing) are likely over. But those who embrace and adapt to the coming changes, and surround themselves with others who have done the same, will flourish.

When major companies, set in their ways, try to convince us that change is bad and that we must stick to the status quo, no matter how much human inquisitiveness and ingenuity try to propel us forward, we must look past them. We must know in our hearts that these changes will come, and that if we welcome them into our world, we'll become more successful, more free, and more full of light than we could have ever possibly imagined.

Ray Kurzweil once wrote, "The Singularity is near." The only question will be whether we are ready for it.

In fact if this sounds astonishing, well it's because it truly is! In an interview with the Sun newspaper Kurzweil maintains, that himself and many other scientists now believe that in around 20 years we will have the means to reprogram our bodies' stone-age software so we can halt, then reverse, ageing. Then Nano-technology will let us live forever.

Already, blood cell-sized submarines called nanobots are being tested in animals. These will soon be used to destroy tumors, unblock clots and perform operations without scars. Ultimately, nanobots will replace blood cells and do their work thousands of times more effectively.

These technologies should not seem at all fanciful. Our phones now perform tasks we wouldn't have dreamed possible 20 years ago. In fact Kurzweil says that when he was a student in 1965, his university's only computer cost $12 million and was huge. Today your mobile phone is a million times less expensive and a thousand times more powerful. That's a billion times more capable for the same price. We will experience another billion-fold increase in technological capability for the same cost in the next 25 years.

He maintains that in 2008 we discovered skin cells can be transformed into the equivalent of embryonic cells. So organs will soon be repaired and eventually grown. In a few years most people will have their entire genetic sequences mapped. Before long, we will all know the diseases we are susceptible to and gene therapies will mean virtually no genetic problems that can't be erased.

He also believes that it's important to ensure we get to take advantage of the upcoming technologies by living well and not getting hit by a bus. By the middle of this century we will have back-up copies of the information in our bodies and brains that make us who we are. Then we really will be immortal.

Personally I believe this is just a bridge to the rise of machines empowering our lives as we merge with machines.

Why must we merge with the machines? Well the machines are getting better exponentially. In fact the software on my smartphone updates every day while our genome hasn't been updated much in 100000 years! Well in that case if you think about it we have software programs in our genome that need to be updated, and I truly believe the 21st century will see breakthroughs in that field of genomics in updating our software.

At this point you ask yourself?
Is the greatest story ever told?

You just got a glimpse into the future based on the law of accelerating returns and the laws of physics do allow us to merge with machines and live in a world with A.I. augmenting our life every day.

Well if that's the implication you might ask well what happens to our jobs if machines are getting smarter?

The common question I always get no matter whenever I bring up the singularity and new technologies is what happens to jobs? Are jobs going to go away?

In fact a very good concerned friend of mine Nyc Labrets on Facebook posted this thread below:

> In a very short 10 years from now, Moore's Law, which dictates that computer processing power doubles *exponentially* every 2 years, while the cost remains constant, will hit "Physical Limits" mainly because, at the rate we're going, we're going to have transistors that are the size of atoms, and it's impossible to get any smaller than that.
>
> However, between now and then, in the year 2022, the average computer will be somewhere around 15 to 30 times more

powerful than they are today: http://www.facebook.com/note.php?note_id=10150156682307871

Even if things plateau at that point it will still have a more dramatic transformative effect on our world than the widespread adoptions of the automobile and airplane did less than 100 years ago.

By the time the year 2022 rolls around a machine like IBM's 'Watson' that just beat 2 human competitors on Jeopardy! will cost all of $30,000 to $15,000 for a company like Wal-Mart to buy. Or less. Given the choice to invest in a frail, unreliable, human worker, or one of these machines, where do you think that Wal-Mart, and every other major company that can afford it, is going to put their money?

Since the ownership of these machines, and the systems that they drive, like Factory & Warehouse Automation, remain in the hands of just a relatively small group of a handful of Owner-Operators, (maybe 10 to 20 million people, tops, worldwide, since they are the only ones that can afford to implement and deploy this tech on a wide-scale), the vast majority of Earth's present population will *not* be invited to share in the fruits that these machines will be providing to those that own them.

One of the Key Assumptions that Ray Kurzweil is making is that within the next 5 to 7 years some sort of Breakthrough is going to come along and extend the life of Moore's Law for at least another 1 to 2 decades longer past its Physical Limits Expiration Date of 2022.

What does that look like?

I now have a 1 year old Niece.

In the year 2032, when she is due to graduate College, if there is no interruption to Moore's Law after the year 2022, then the *average computer will have processing power that is roughly 1,000 times more powerful than today's computers.*

If she continues on to pursue an Advanced Degree after getting her BA, by the time she graduates Law or B School, computers will be almost 5,000 times as powerful as today machines, and if she decides to become a Doctor, then by the time she finishes her Residency and is able to hang out her Shingle as Licensed MD, computers will be 10,000 times as powerful as they are now.

All of this begs the question, why on earth should she be going to College in the first friggin' place?

It would surely appear that way, I mean the more advanced technology gets technology replaces humans by doing a more efficient job than humans in many things.

Thus, we end up with charts that look like this productivity:

According to Barry Lawrence, director of the Global Supply Chain Laboratory at Texas A&M University, he maintains "The argument that they're taking American jobs is a bit overblown," in reference to all the mainstream media talk on manufacturing and creating jobs.

For instance, consider the fact that "The biggest job drain to manufacturing is actually automation," he says. Simply put, with production advancements and process improvements, manufacturers are now capable of producing more product with fewer people.

According to figures presented in a recent article from the Wall Street Journal, manufacturing productivity has increased 103 percent since the late 1980s. We're still making things; we just need fewer bodies to actually do it.

An analysis by Paul Weener, managing partner at IntelliQ Research in State College, PA, that ran in the Centre Daily Times compares the shift occurring now to the shift that has already occurred in agriculture:

"In 1900, 44 percent of all jobs were in agriculture; today 2.4 percent of jobs are in agriculture, but we produce a lot more food now than we did in 1900. Likewise, we are producing a lot more manufactured goods than we did 40 years ago with fewer employees."

Weener goes on to ask: "Should we try to restore those millions of agriculture jobs that have been lost? Should we try to restore those 350,000 switchboard operator jobs lost since 1970?"

In other words, manufacturing may not be the solution to the unemployment problem we have in the U.S. But that doesn't mean manufacturing isn't strong: We just might have to reconsider how we measure its strength.

Consider one example, agriculture for instance, people always say, "hey if we all become scientists and smart and artists? Who shall grow our food?" Well precisely, as controversial as this sounds, I will say that technology, especially these new exponential technologies are deeply democratizing and bring down the costs of resources globally. For instance the dense metropolis of Singapore is now home to the world's first commercial vertical farm! Built by Sky Greens Farms, the rising steel structure will help the city grow more food locally, reducing dependence on imported produce. The new farm is able to produce 1 ton of fresh veggies every other day, which are sold in local supermarkets.

The world's first commercial vertical farm will provide a fresh new source of sustainable produce for Singaporeans. The tiny country currently produces only 7% of its vegetables locally, driving a need to buy from other countries. But thanks to the new vertical farm, citizens can eat locally produced food.

The farm itself is made up of 120 aluminum towers that stretch thirty feet tall. Looking like giant greenhouses, the rows of plants produce about a half ton of veggies per day. Only three kinds of vegetables are grown there, but locals hope to expand the farm to include other varieties. The farm is currently seeking investors to help build 300 additional towers, which would produce two tons of vegetables per day. Although the $21 million dollar price tag is hefty, it could mean agricultural independence for the area.

The vertical farm veggies have become a big hit with the locals too. Although the produce costs 10 to 20 cents more than other veggies at the supermarket, consumers seemed eager to buy the freshest food possible – often buying out the market's stock of vertical farm foods. This innovative vertical farm could help change the way the world eats, giving dense cities an opportunity to grow food in their own back yard.

Source: http://inspirationgreen.com/vertical-farms.html

The advantage of vertical farms is that they can feed billions of people require very little land space, and don't require deforesting lands. In fact since over 3 billion of our people will live in cities soon, more and more of these should start popping up all over town in major cities. You can save tons of land from being, deforested, grow food locally and make it organic, not requiring pesticides. Also we don't have to transport food long distances. Since societies that are technologically based have 427:1 productivity per capita between the least and most poor versus agricultural societies which have a 5:1 ratio between the most and least productive, I imagine all these technologically literate towns like San Francisco, Hong Kong, and Singapore and even desert nations can grow these and water plants using recycled water or gray

water.

The Singularity Culture goes mainstream

The culture of the singularity is starting to go way mainstream nowadays. How much more mainstream do you get than the starting of an entire university devoted to the study of AI, Futurism, and many more, in fact the program concentrates on six exponential growing technologies areas which include:

1. Artificial Intelligence & Robotics
2. Nanotechnology
3. Biotechnology & Bioinformatics
4. Medicine & Neuroscience
5. Networks & Computing Systems
6. Energy & Environmental Systems

We see early examples of AI, on our phones today right? To me the cell phone represents the ultimate mental prosthesis we have today. What other device merges man and machine to all of human information I mean look at the number of options of A.I. on our phones that handle complicated queries. There is Samsung S voice, Siri, Google Now,

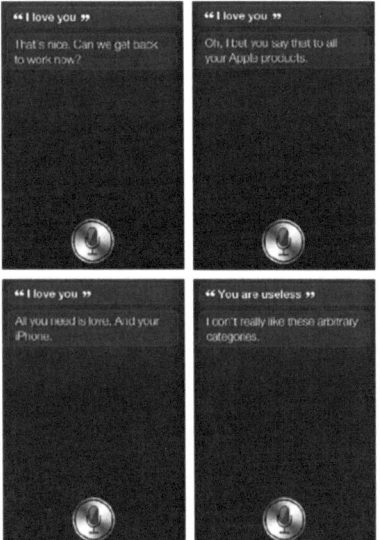

Microsoft ask Ziggy, and so many more.

The most popular assistant that started taking AI mainstream would be Siri. In fact Siri has its origins in a research project begun in 2003 and funded by the U.S. military's Defense Advanced Research Projects Agency (DARPA). The effort was led by SRI International, which in 2007 spun off a company that released the original version of Siri as an iPhone app in February 2010. When I first tried it on my iPhone I loved the ability to ask natural language queries and the fact it linked complicated queries to Wolfram Alpha a search engine that organizes facts and gives answers using computation over just links. This makes the question and answer much more personal and gives answers to questions in natural language like magic. According to the magazine Technology Review they maintain that you can ask Siri things like reminding you to wake up at 8:00 a.m. and it will set the phone's alarm clock accordingly. Tell Siri to send a text message to a friend and it will dutifully take dictation before firing off your missive. Say "Where can I find a burrito, Siri?" and Siri will serve up a list of well-reviewed nearby Mexican restaurants, found by querying the phone's location sensor and performing a Web and map search. Siri also has countless facts and figures at its fingertips, thanks to the online "answer engine" Wolfram Alpha, which has access to many databases. Ask "What's the radius of Jupiter?" and Siri will casually inform you that it's 42,982 miles. In fact as the queries in the image in the preceding page show Siri even answers basic nonsense queries and jokes. However this was just the start. Next let's look at the future of deeper A.I.

IBM Watson on Jeopardy

"The design of the human brain, while not simple, is nonetheless a billion times simpler than it appears, due to massive redundancy. Biological systems such as the brain are probabilistic fractals, which give them their messy, unpredictable quality. The design of the human brain is in the genome, and I show that there are only about 50 million bytes of design information (after lossless compression) in the genome (including the epigenetic information in the reproductive machinery), which is a level of complexity we can handle. And indeed, we are making exponential gains in modeling and simulating extensive regions of the human brain, including the cerebral cortex." — Ray Kurzweil

The singularity and the power of Artificial intelligence going mainstream is no more evident than in the IBM Supercomputer named Watson. In fact I've been passionately following IBM's supercomputer development for a vast number of years. IBM originally began developing Watson to play jeopardy as a machine capable of deep question and answering capabilities. Although Jeopardy seems like an easy enough game, it's actually a very tricky game, Watson has to understand jokes, puns, nuances of language, metaphors, similes, various expressions. In fact it was so good that it beat the best two Jeopardy players at the game combined!

Source: http://www.bloomberg.com/news/2012-08-28/ibm-creating-pocket-sized-watson-in-16-billion-sales-push-tech.html

I mean IBM did a good job playing Jeopardy, however, let's face it's still not the Artificial intelligence that is the universal knowledge engine we've dreamed of yet.

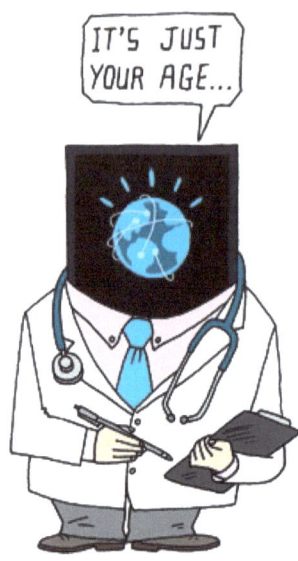

The universal answer engine we've dreamed of in our pocket.

Source: Bloomberg Businessweek

According to Matt Ridley, Last year the IBM program, Watson, triumphed the tough game of natural language by addressing tough challenges by winning "Jeopardy!" (Sample achievement: Watson worked out that a long, tiresome speech delivered by a frothy pie topping was a "meringue harangue.")

According to Sebastian Anthony of the website Extreme Tech, Watson is an artificial intelligence that is capable of answering very complex questions using natural language answers. In essence, IBM is hoping to build a better, faster, and more professional/enterprisey version of Apple's Siri, the voice-controlled assistant that debuted on the iPhone 4S. Imagine your brain or your Google glasses hooked up to this app right? Like you would be able to ask any complicated query with metaphors, puns, similes and even medical questions and get a list of good answers. In fact the more you ask and more it learns it will get smarter. In essence we all will have access to the ultimate star trek computer for free soon.

In fact now, the refrigerator-size supercomputer machine is prepping for a second career as part of a trial program at New York City's Memorial Sloan-Kettering Cancer Center. Watson relies on parallel processing -- geek speak for running multiple tasks at once -- to sift through 500 gigabytes of data per second. A physician can enter the results of a biopsy, for example, and Watson pulls relevant bits of a patient's history as well as clinical studies and medical journals. It then lists potential diagnoses and their varying "levels of confidence," or probability. The final call is left up to the doctor. The idea is to put the powerful supercomputer that can read 100 years of medical journals and give a statistical diagnosis and multiple different diagnosis and let a physician evaluate his or her prediction. This will soon also be available in an app that IBM plans to work on. According to Anthony, he states that in theory, Watson's question answering ability would utterly blow Siri and Google Now out of the water. While Siri can set your alarms, Watson can parse a patient's charts and provide clinical diagnoses and pharmaceutical prescriptions. Where Siri can tell you whether you'll need an umbrella, you could ask Watson whether now is the right time to plant your crops — or for a complete walkthrough on how to fix your toaster. This will essentially put a doctor in the palm of every person that has a smartphone all 4 billion of us in the next few years!

Source: http://www.extremetech.com/computing/135173-ibm-working-on-watson-app-for-smartphones

This is just the start in emulation artificial intelligence. I predict we are going exponentially further soon indeed. In fact the inventor and futurist Ray Kurzweil in his new book "How to Create a Mind." Mr. Kurzweil reckons that a full understanding and simulation of the human brain is a lot closer than most people think. Since he has a more impressive track record of predicting technological progress than most, so pay attention.

So how close are we to replicating AI? According to Ray, for a start, the brain is built from a relatively small and simple body of information—the 25 million bytes of the genome. The complexity comes from ordered growth and elaboration. Second, the brain contains massive redundancy, with certain kinds of basic pattern-recognizing circuits repeated maybe 300 million times in different brain regions. Third, as Van Wedeen of Harvard Medical School and colleagues found in a recent study, much of the brain has a horizontal grid of fibers running at right angles, connecting vertically: a bit like the streets and elevators of the city of Manhattan.

Maybe Watson's Deep language understanding and looking for and understanding information will get better if it learns to learn associations and patterns and its "brain" becomes more like ours.

However, this project is just the start, because IBM, aka big blue is working on a project called blue brain which aims to reverse engineer the brain.

What is Blue Brain? Although direct brain emulation using artificial neural networks on a high-performance computing engine is a common approach being employed by IBM.

Since November 2008, IBM received a $4.9 million grant from the Pentagon for research into creating intelligent computers. The Blue Brain project is being conducted with the assistance of IBM in Lausanne. The project is based on the premise that it is possible to artificially link the neurons "in the computer" by placing thirty million synapses in their proper three-dimensional position.

In March 2008, *Blue Brain* project was progressing faster than expected: "Consciousness is just a massive amount of information being exchanged by trillions of brain cells.[9]" Some proponents of strong AI speculate that computers in connection with Blue Brain and Soul Catcher may exceed human intellectual capacity by around 2015, and that it is likely that we will be able to download the human brain at sometime around 2050.

In fact IBM seems pretty ambitious recently with all these new projects and research in deep AI and next generation of machine learning. In fact IBM claims Within the next five years, advances in cognitive computing will let humans use computers to replicate sensory experiences in a virtual setting, as well as to augment and expand sensory experiences beyond human capabilities, IBM has predicted. They reflect "a lot of things going on in the laboratory for the year, or in some cases several years," IBM CTO of Telecom Research Paul Bloom told the internet website TechNewsWorld. According to IBM, In five years, technology advancements could enable sensors to analyze odors or the molecules in a person's breath to help diagnose diseases. Imagine, the applications of the proposed "Qualcomm Tricorder X prize" under development, IBM might be the company with the technology to power such a device that can detect sensory information and replacing a doctor for real.

According to Paul Bloom advances In sight technologies will give computers superhuman vision. Computers see images, photographs and video as pixels, and they rely on tags to determine what is in an image. IBM and other companies are developing systems that let computers analyze images pixel by pixel.

In medical applications, computers will be able to analyze medical imaging technologies from MRIs, CT scans, X-rays and ultrasounds to identify tumors, for example. This technology could allow computers to detect problems that humans aren't able to discern in standard images.

"A system will be able to understand and read an X-ray and be able to see things that a human can't," Bloom said. "A system like this could be more accurate than the capabilities of a human being."

What makes me really optimistic is the article I've basically summarized below by Dean Takahashi, a writer for Venture beat discussing IBM and what they're doing with chips that can emulate the way the brain processes human information, entitled, "IBM produces first working chips modeled on the human brain

Dean writes that Big Blue recently announced that it, along with four universities and the Defense Advanced Research Projects Agency (DARPA), have created the basic design of an experimental computer chip that emulates the way the brain processes information.

He moreover, maintains that IBM's so-called cognitive computing chips could one day simulate and emulate the brain's ability to sense, perceive, interact and recognize — all tasks that humans can currently do much better than computers can.

Dharmendra Modha pictured below, is the principal investigator of the DARPA project, called Synapse (Systems of Neuromorphic Adaptive

Plastic Scalable Electronics, or SyNAPSE). He is also a researcher at the IBM Almaden Research Center in San Jose, Calif.

"This is the seed for a new generation of computers, using a combination of supercomputing, neuroscience, and nanotechnology," Modha said in an interview with VentureBeat. "The computers we have today are more like calculators. We want to make something like the brain. It is a sharp departure from the past."

According to Dean he writes that "If it eventually leads to commercial brain-like chips, the project could turn computing on its head, overturning the conventional style of computing that has ruled since the dawn of the information age and replacing it with something that is much more like a thinking artificial brain. The eventual applications could have a huge impact on business, science and government. The idea is to create computers that are better at handling real-world sensory problems than today's computers can. IBM could also build a better Watson, the computer that became the world champion at the game show Jeopardy earlier this year."

Source: http://venturebeat.com/2011/08/17/ibm-cognitive-computing-chips/

Dean says that now that the researchers have completed phase one of the project, which was to design a fundamental computing unit that could be replicated over and over to form the building blocks of an actual brain-like computer.

Richard Doherty, an analyst at the Envisioneering Group, has been briefed on the project and he said there is "nothing even close" to the level of sophistication in cognitive computing as this project.

This new computing unit, or core, is analogous to the brain. It has "neurons," or digital processors that compute information. It has "synapses" which are the foundation of learning and memory. And it has "axons," or data pathways that connect the tissue of the computer.

Source: http://venturebeat.com/2011/08/17/ibm-cognitive-computing-chips/

According to Dean, he writes that," while it sounds simple enough, the computing unit is radically different from the way most computers operate today. Modern computers are based on the von Neumann architecture, named after computing pioneer John von Neumann and his work from the 1940s.

In von Neumann machines, memory and processor are separated and linked via a data pathway known as a bus. Over the past 65 years, von Neumann machines have gotten faster by sending more and more data at higher speeds across the bus, as processor and memory interact. But the speed of a computer is often limited by the capacity of that bus, leading some computer scientists to call it the 'von Neumann bottleneck'."

Dean says, "The brain-like processors with integrated memory don't operate fast at all, sending data at a mere 10 hertz, or far slower than the 5 gigahertz computer processors of today. But the human brain does an awful lot of work in parallel, sending signals out in all directions and getting the brain's neurons to work simultaneously. Because the brain has more than 10 billion neuron and 10 trillion connections (synapses) between those neurons, that amounts to an enormous amount of computing power.
IBM wants to emulate that architecture with its new chips."

"We are now doing a new architecture," Modha said. "It departs from von Neumann in variety of ways."

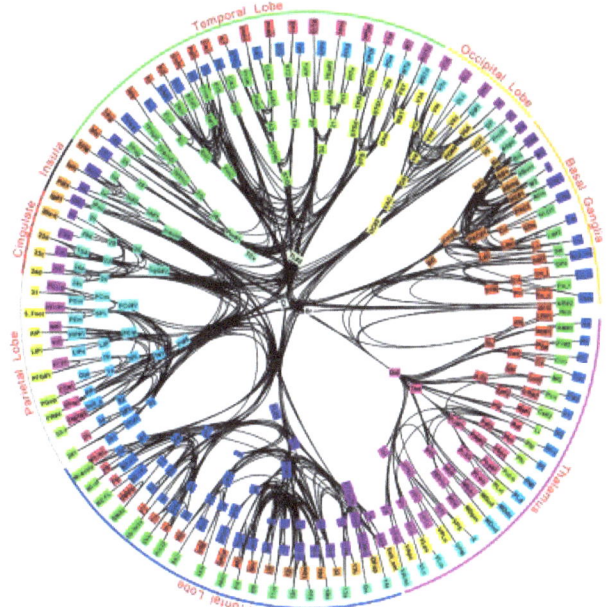

Source: http://venturebeat.com/2011/08/17/ibm-cognitive-computing-chips/

Dean maintains that the research team has built its first brain-like computing units, with 256 neurons, an array of 256 by 256 (or a total of 65,536) synapses, and 256 axons. (A second chip had 262,144 synapses) In other words, it has the basic building block of processor, memory, and communications. This unit, or core, can be built with just a few million transistors (some of today's fastest microchips can be built with billions of transistors).

Modha said that this new kind of computing will likely complement, rather than replace, von Neumann machines, which have become good at solving problems involving math, serial processing, and business computations. The disadvantage is that those machines aren't scaling up to handle big problems well any more. They are using too much power and are harder to program.

According to Modha, the more powerful a computer gets, the more power it consumes, and manufacturing requires extremely precise and expensive technologies. And the more components are crammed together onto a single chip, the more they "leak" power, even in stand-by mode. So they are not so easily turned off to save power.

The advantage of the human brain is that it operates on very low power and it can essentially turn off parts of the brain when they aren't in use.

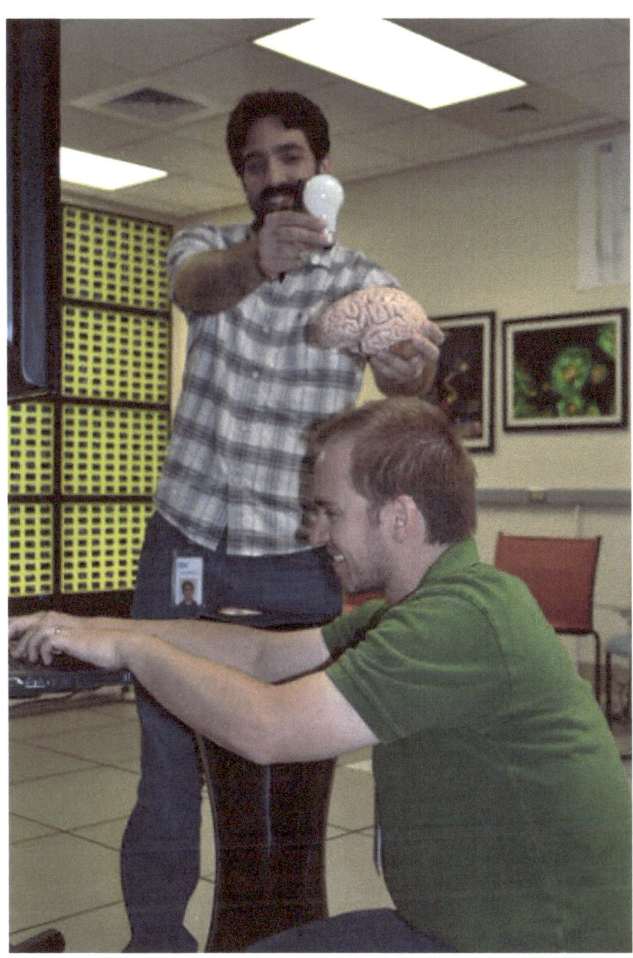

Source: http://venturebeat.com/2011/08/17/ibm-cognitive-computing-chips/

Furthermore, Dean writes "these new chips won't be programmed in the traditional way. Cognitive computers are expected to learn through experiences, find correlations, create hypotheses, remember, and learn from the outcomes. They mimic the brain's "structural and synaptic plasticity." The processing is distributed and parallel, not centralized and serial.
With no set programming, the computing cores that the researchers have built can mimic the event-driven brain, which wakes up to perform a task.

Modha said the cognitive chips could get by with far less power consumption than conventional chips.

The so-called "neurosynaptic computing chips" recreate a phenomenon known in the brain as a "spiking" between neurons and synapses. The system can handle complex tasks such as playing a game of Pong, the original computer game from Atari, Modha said.

Two prototype chips have already been fabricated and are being tested. Now the researchers are about to embark on phase two, where they will build a computer. The goal is to create a computer that not only analyzes complex information from multiple senses at once, but also dynamically rewires itself as it interacts with the environment, learning from what happens around it.

What's more astonishing according to Dean is that the chips themselves have no actual biological pieces. They are fabricated from digital silicon circuits that are inspired by neurobiology. The technology uses 45-nanometer silicon-on-insulator complementary metal oxide semiconductors. In other words, it uses a very conventional chip manufacturing process. One of the cores contains 262,144 programmable synapses, while the other contains 65,536 learning synapses.

Furthermore, he writes "Besides playing Pong, the IBM team has tested the chip on solving problems related to navigation, machine vision, pattern recognition, associative memory (where you remember one thing that goes with another thing) and classification."

Eventually, Dean Takahashi believes that IBM will combine the cores into a full integrated system of hardware and software. IBM wants to build a computer with 10 billion neurons and 100 trillion synapses, Modha said. That's as powerful as the human brain. The complete system will consume one kilowatt of power and will occupy less than two liters of volume (the size of our brains), Modha predicts. By comparison, today's fastest IBM supercomputer, Blue Gene, has 147,456 processors, more than 144 terabytes of memory, occupies a huge, air-conditioned cabinet, and consumes more than 2 megawatts of power.

As a hypothetical application, IBM said that a cognitive computer could monitor the world's water supply via a network of sensors and tiny motors that constantly record and report data such as temperature, pressure, wave height, acoustics, and ocean tide. It could then issue tsunami warnings in case of an earthquake. Or, a grocer stocking shelves could use an instrumented glove that monitors sights, smells, texture and temperature to flag contaminated produce. Or a computer could absorb data and flag unsafe intersections that are prone to traffic accidents. Those tasks are too hard for traditional computers.

Synapse is funded with a $21 million grant from DARPA, and it involve six IBM labs, four universities (Cornell, the University of Wisconsin, University of California at Merced, and Columbia) and a number of government researchers.

Dean maintains that for phase 2, IBM is working with a team of researchers that includes Columbia University; Cornell University; University of California, Merced; and University of Wisconsin, Madison. While this project is new, IBM has been studying brain-like computing as far back as 1956, when it created the world's first (512 neuron) brain simulation.

"If this works, this is not just a 5 percent leap," Modha said. "This is a leap of orders of magnitude forward. We have already overcome huge conceptual roadblocks."

According to me there are good reasons to believe that, regardless of implementation strategy, the predictions of realizing artificial brains in the near future are optimistic.

Enter the Google Car

Probably the most impressive feat of AI and one that will come true in the next 5-10 years in all cars as an option and start of on high end cars first is the self-driving vehicle. In fact this whole feat started out as a **DARPA (Defense Advanced Research Project Agency)** competition many years ago. There emerged a winning car. The goal was to make a car drive autonomously from a starting point to a finish line on its own. Then in 2007 came the **DARPA** urban challenge which was a competition for a car to drive through an urban environment. They started out looking like the car shown here:

Source: http://www.google.com/imgres?imgurl=http://low-powerdesign.com/sleibson/wp-content/uploads/2011/04/GM-Boss.jpg

However in the last few years the technology has exponentially advanced. I mean hugely there are two teams I'm familiar with well that are doing amazing work, one in Berlin and the other Google is testing a fleet of self-driving cars in California under the worst conditions, through traffic, day and night, intersections, and even crooked Lombard Street in San Francisco. The cars drive themselves through all these amazing environments.

As of this book California, Nevada and some other states are in the process of legalizing self-driving cars. In fact the Google team has equipped a test fleet of at least eight vehicles, consisting of six Toyota Prius, an Audi TT, and a Lexus RX450h, each accompanied in the driver's seat by one of a dozen drivers with unblemished driving records and in the passenger seat by one of Google's engineers. The car has traversed San Francisco's Lombard Street, famed for its steep hairpin turns and through city traffic. The vehicles have driven over the Golden Gate Bridge and on the Pacific Coast Highway, and have circled Lake Tahoe. The system drives at the speed limit it has stored on its maps and maintains its distance from other vehicles using its system of sensors. The system provides an override that allows a human driver to take control of the car by stepping on the brake or turning the wheel, similar to cruise control systems already in cars.

Source: http://mashable.com/2012/08/07/google-driverless-cars-safer-than-you/

This is what the latest version of the Google car looks like. This one was made to be tested in all weather systems. According to Mashable, how safe a driver is your average robot? Safer than your average American, at least by one measure they say.

Google announced that its self-driving cars have completed 300,000 miles of test-drives, under a "wide range of conditions," all without any kind of accident. To put that into perspective, the average U.S. driver has one accident roughly every 165,000 miles. Here's how they got that figure: our average mileage per year is 16,550, according to the Federal Highway Administration; the average length of time we go between traffic accidents is 10 years, according to Allstate. The Google project uses mainly Toyota Priuses equipped with a range of cameras, radar sensors and laser range-finders to see other traffic and the sophisticated software uses Google Maps to navigate routes. As you can see the technology using Google's backend has seriously advanced in just a few years of research. According to the Google Co-founder Sergey Brin, he maintains that in just 5 years we will have Google cars for the average American to be used as taxi cabs.

The project is the brainchild of Sebastian Thrun, the 43-year-old director of the Stanford Artificial Intelligence Laboratory, a Google engineer and the co-inventor of the Street View mapping service.

In 2005, he led a team of Stanford students and faculty members in designing the Stanley robot car, winning the second Grand Challenge of the Defense Advanced Research Projects Agency, a $2 million Pentagon prize for driving autonomously over 132 miles in the desert.

Besides the team of 15 engineers working on the current project, Google hired more than a dozen people, each with a spotless driving record, to sit in the driver's seat, paying $15 an hour or more. Google is using more than six Priuses and an Audi TT in the project.

The Google researchers said the company did not yet have a clear plan to create a business from the experiments. However, Dr. Thrun is known as a passionate promoter of the potential to use robotic vehicles to make highways safer and lower the nation's energy costs. It is a commitment shared by Larry Page, Google's co-founder, who appears to be a backer of all things tech and cutting edge or innovative.

Education of the Future

In fact more astonishing is how technology is replacing and disrupting the entire education industry. On the forefront is Dr. Sebastian Thrun, who's been heading the Google Car project actually with his other venture Udacity! More on that in a bit.

According to Qualcomm's Spark blog, they maintain that the industrial-era model of education—one teacher lecturing to students for a set period of time using a narrow set of resources —is no longer how the real world works. But it is the education model that persists today in schools, and it is critical that we transform the current state of education.

Moreover, I totally agree with, them. In fact, legendary Silicon Valley figure Peter Thiel, first investor in Facebook and founder of Clarium Capital maintains that higher education is facing a big bubble right now. In fact, a bubble is when something is overvalued by hype and faces very little in returns. Consider that the cost of a college education costs up to 45000 dollars a year for a piece of paper and in return you get very little in terms of credentials. Maybe a few decades ago, this would have been worth it, however in today's modern age, traditional colleges do very little to prepare us for the world of today and tomorrow. For instance, most of your success depends on your ability to network and understand a niche of skills and learn on the job. In fact really good social networking can be done by living in a good city and attending many valuable events and university events for free. To pick up skills, almost every course and skill is available in community colleges, or trade schools or online. In fact considering the total cost of a university education nowadays, it actually makes more sense to save that money and invest in a business and learn on your own time instead of wasting four years. Of course attending a top school like MIT, Harvard or Stanford is still very much worth it for the connections you make, the alumni networks and of course the amount of prestigious professors you make first hand connections with, but attending a lonely liberal arts four year college has almost no returns or connections and is a very poor investment. That's precisely why higher education is in the midst of an enormous bubble according to me as well. Essentially the cost of a higher education is overvalued and technology is growing exponentially and getting credentials nowadays for high paying skills could not be easler or cheaper, essentially anyone with an internet connection can get the best education in the world for free and many courses are becoming certified online. Moreover, anyone who takes online education and completes coursework probably took a course to truly learn and not cheat their way through it. With the plethora of options available online, it makes sense that students will find and take mostly likely what they're most passionate about studying and learning.

In fact there are many initiatives that are disrupting education. For instance, Khan Academy is a great supplement for anyone in school or wants to learn the basics on any subject with its set of YouTube videos. Moreover, the website now also have self-help tests and tutorials to help you get a good education or learn more on a topic of your interest on your own.

However, the real revolution that's happening is in higher education in fact for most high paying jobs no longer is a college education needed. If you want to be a lawyer or doctor for now you need credentialing, but for almost every computer science course or engineering almost all can be done online for free to get your credentials and in fact the online courses are far superior to the college ones in many ways. For instance, the best colleges offer classes on Coursera and EdX and many new colleges are joining every month. I was reading an article on the technology website GigaOM recently and they maintained how Anant Agarwal, director of MIT's Computer Science and Artificial Intelligence Laboratory and a driving force behind edX said the technology could educate up to 1 billion people. "Anyone with an Internet connection anywhere in the world can have access [to edX]," said Harvard President Drew Faust.

Moreover, I was reading an article on how the schools of the future will need to adapt to the online model much like the news print industry or face extinction. Why would someone pay 45000 dollars a year for an education anyone in the world can obtain on an iPad with a 2 dollar a month 3G internet connections (Yes, in India that's the cost of the cheapest unlimited 3G plans). Moreover, according to a recent New York Times article, the co-founders of the popular online University Coursera, computer science professors at Stanford University, watched with amazement as enrollment passed two million last month, with 70,000 new students a week signing up for over 200 courses, including Human-Computer Interaction, Songwriting and Gamification, taught by faculty members at the company's partners, 33 elite universities. So the best education in the world for all high school students and undergraduate degrees is already quite ubiquitous today to anyone with a smartphone, tablet or PC device.

Enter the Surrogates

"You are about to become obsolete. You think you are special, unique and that whatever it is that you are doing is impossible to replace. You are wrong." This is the start to a great book I read this year by Federico Pistano, author of the amazing book, "Robots will steal your job, but that's okay: How to survive the economic collapse and be happy." In fact I would recommend this book to anyone who's interested in the learning what's going to happen. One such example would be in the field of medicine. Enter Vinod Khosla, the former founder of Sun Microsystems and legendary Silicon Valley investor maintains that Technology will replace 80% of what doctor's do, that's right. Why? Well you see over the last few years there has been an exponential trend in AI systems and humanoid robots that was devoid just a decade ago. So many startups and the Silicon Valley titans are investing their cash into these moonshot projects. The idea is to disrupt fields that have been areas of great stagnation and I truly feel it's inevitable. For instance, Vinod Khosla maintains that for 150 years, doctors have routinely prescribed antipyretics like ibuprofen to help reduce fever. But in 2005, researchers at the University of Miami, Florida, ran a study of 82 intensive care patients. The patients were randomly assigned to receive antipyretics either if their temperature rose beyond 101.3°F ("standard treatment") or only if their temperature reached 104°F. As the trial progressed, seven people getting the standard treatment died, while there was only one death in the group of patients allowed to have a higher fever. At this point, the trial was stopped because the team felt it would be unethical to allow any more patients to get the *standard treatment*. Therefore, Khosla asks the question, "when something as basic as fever reduction is a hallmark of the "practice of medicine" and hasn't been challenged for 100+ years, we have to ask: What else might be practiced due to tradition rather than science?"

In fact according to Khosla, healthcare should become more about data-driven deduction and less about trial-and-error. He believes that's hard to pull off without technology, because of the increasing amount of data and research available. Next-generation medicine will utilize more complex models of physiology, and more sensor data than a human MD could comprehend, to suggest personalized diagnosis. Thousands of baseline and multi-omic data points, more integrative history, and demeanor will inform each diagnosis.

In a new article to Fortune magazine Khosla goes on to say that much of what physicians do (checkups, testing, diagnosis, prescription, behavior modification, etc.) can be done better by sensors, passive and active data collection, and analytics. But, he says doctors aren't supposed to just measure. They're supposed to consume all that data, consider it in context of the latest medical findings and the patient's history, and figure out if something's wrong. Computers can take on much of that diagnosis and treatment and even do these functions better than the average doctor (while considering more options and making fewer errors). Most doctors couldn't possibly read and digest all of the latest 5,000 research articles on heart disease. And, most of the average doctor's medical knowledge is from when they were in medical school, while cognitive limitations prevent them from remembering the 10,000+ diseases humans can get.

Computers are better at organizing and recalling complex information than a hotshot Harvard MD. They're also better at integrating and balancing considerations of patient symptoms, history, demeanor, environmental factors, and population management guidelines than the average physician. Besides, 50% of MDs are below average! Computers also have much lower error rates. Therefore, he begs the question, shouldn't we take advantage of that when it comes to our health?!

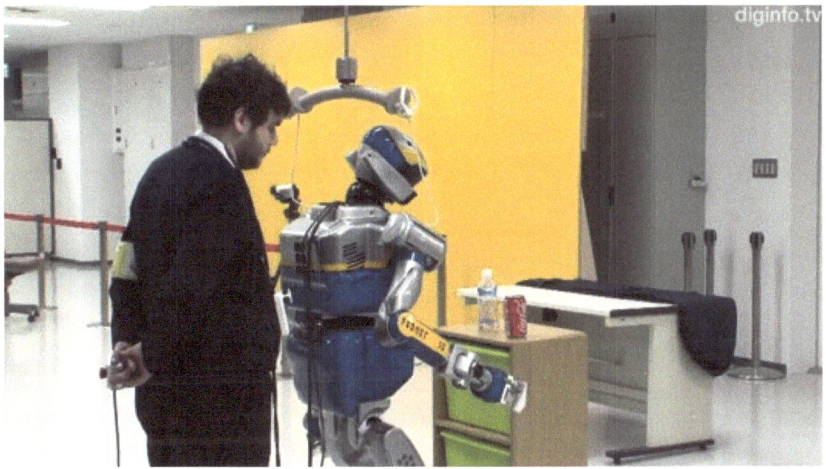

Source: http://www.gizmag.com/mind-controlled-robot-avatars/24994/

Researchers at the CNRS-AIST Joint Robotics Laboratory (a collaboration between France's Centre National de la Recherche Scientifique and Japan's National Institute of Advanced Industrial Science and Technology) are developing software that allows a person to drive a robot with their thoughts alone. The technology could one day give a paralyzed patient greater autonomy through a robotic agent or avatar. In fact this is the idea behind a project called 2045 in the initial stages and even DARPA, to have robots that are mind controlled. In fact this has been an idea of mine to build for a while. I used to work in a Brain Machine interface lab that actually uses this technology called the P300 EEG system from which a user can type text using thoughts alone or even operate a robotic arm, with some good programming and machine intelligence combined, it seems logical the next step would be a robotic system to help the elderly. If this seems like something from science fiction, well it's because it is. If you've seen the movie Surrogates with Bruce Willis, they did a futuristic version of what's already being done today in the early stages. Mind Controlled surrogate avatars are coming very soon indeed, in fact it's because we are working on them. I have a friend who's programming the robotic steps which is more complicated and I've worked in a lab where we were using the brain machine interface to control anything from a light switch to changing television channels. If this is what's possible now its far off to imagine that soon, I will be able think of beats or music in my head and they will appear recorded real time in computers. Perhaps even in the future I may be able to imagine any voice in my head and let it appear in the computer perfectly played out!

Source: http://www.gizmag.com/mind-controlled-robot-avatars/24994

According to the website *Transparency Revolution*, most people agree that machines are rapidly taking on more and more tasks that at one time could only be performed by humans, not everyone is convinced that automation represents a real threat to the long-term employability of human beings — as long we're willing to keep moving up the value chain and do things machines can't.

Some will argue that, that based on history, only luddites and the technologically (or economically) ignorant fear a machine takeover of the full human employment space. The "threat" of automation has never panned out. So far. Then again, machines are getting smarter and faster at an exponential rate; we are not. So it could be that we will all end up doing only creative things like making music and movies and art, thus experiencing a transition to a whole new economic model, one wherein we are increasingly employed in the lucrative field of value-added intermediation — filling the gaps the machines can't.

On the other hand, machine capabilities may improve so fast that they leave very little gap for us to intermediate. Or there might be limited economic value in the gaps we can fill. The possibility of a truly painful transition is out there and it's very real.

In the long run, however, there is an exponential trend that favors us. The overall human condition is improving exponentially, and has been for some time. Vast improvements in machine capability are just a subset of, and contributing factor towards, that overall trend. In fact, it may turn out that we need these machines — and their ability to do everything we can do faster and better — to keep that trend going.

I believe thought controlled avatars will take over the next few years. IEEE magazine maintains that In the movie *Avatar*, humans hooked themselves up to brain-machine-interface pods with which they could control giant genetically engineered human-alien hybrids. It's just a movie, but DARPA, the U.S. Defense Advanced Research Projects Agency, doesn't care: It wants this kind of system to be real, just replace "giant genetically engineered human-alien hybrids" with "robots."

In fact according to IEEE magazine, they say that in 2013 DARPA has decided to pour US $7 million into the "Avatar Project," whose goal is the following: *"develop interfaces and algorithms to enable a soldier to effectively partner with a semi-autonomous bi-pedal machine and allow it to act as the soldier's surrogate."*

That word "surrogate" implies something more than just telepresence, and indeed DARPA does specify that it is looking for *"key advancements in telepresence and remote operation of a ground system."* I believe a surrogate will take over and Evan Ackerman, the author of the article on DARPA maintains that according to me, the implication is that there's going to be some technology that effectively puts the user "inside" the remote system, whether it's through immersive VR or exoskeleton or some sort of direct brain control. Either of these things is a realistic possibility, especially if DARPA's tossing a couple million at the problem.

Humanity enters a new saga

LIVING FOREVER

"Death is a tragedy but the philosophies and religions that developed in pre-scientific times have rationalized it as a 'good thing', since we have had no alternative. But disease and death reflect a great loss of knowledge, of relationships, and of the potential to expand the human experience. Expanding human knowledge is, in my view, the purpose of human life, and staying healthy is a prerequisite for being able to do this. In my mind, people who ignore their health through unhealthy lifestyles and then become a burden to their loved ones and the rest of society are the ones being selfish and short-sighted." — Ray Kurzweil

In fact Ray Kurzweil, has a conversation with a blogger on his website, we shall call blogger X for anonymity sake.

Blogger X asks:

How do you find motivation to want to live forever? How do you find comfort in your father's death, knowing you may never truly see him again — only an avatar of what he'd represent?

Kurzweil responds:
I have the motivation to live to tomorrow, metaphorically speaking. I think everyone has that motivation. As we get to times in the future, we'll have more powerful tools to get to the next stage. As for the death of my father or anyone else, I don't find it comforting, but rather I consider it a tragedy.

Blogger X responds:
I think it's probably true to say, from my limited understanding that it is in our biology to want to survive. We are probably naturally motivated to want to live as long as we can.

The future will probably be fantastic for people who have never experienced loss — but what about those of us who have? We will be in paradise — but what about our deceased family/friends?

How can we fully enjoy a eternal future when there will always be that human part of us that misses our loved ones?

Kurzweil retorts:

If I were to create an avatar of my father (using superintelligent AI to help me do it) based on all of the information I have about him and my and others' memories of him, that avatar would be more like my father was (at age 58, which is how I remember him) than he would be today (at age 100), had he lived.

Ray,

To which Blogger X says:

"It is not demeaning to regard a person as a profound pattern (a form of knowledge), which is lost when he or she dies."

I agree and think it possible that future AI could produce a replica that is a 99.9% clone of that person. At what stage of AI will you be satisfied with the recreation of your father?

"Death is a tragedy."

It is a tragedy in that we seemingly no longer get to experience the universe, and our loved ones are left to experience it without us.

To try and convey my idea, consider this somewhat petty argument: if you're going to be recreated in the future, why do you bother 'fighting' death? Not just because you don't want to leave your loved ones, you enjoy living – but maybe because their is an uncertainty that there is and will only be one Ray?

We can prevent death — do you think it possible that future AI could undo it entirely? Has our understanding of time not been apart of our evolutionary process?

To which Ray Responds:

These are some good insights and a revealing thought experiment.

I would accept a mental clone of my father if it passes a "Fredric Kurzweil Turing test," that is when I cannot distinguish it from my father. It is a somewhat unfair test, however, in that my biological father is not here to compare to, and my memories of him have faded.

Your thought experiment reveals that we are not confident that when a future AI based avatar does pass a Fredric Kurzweil Turing test that it will represent a continuation of his identity. It will be better to scan our brains while we're still here. My preferred scenario is to merge with AI and over time the AI portion of me will become dominant. It will be backed up and it will ultimately understand the remaining biological portion well enough to back that up as well.

Best,

Ray

Moreover, when Reporter Anthony Willams of the Huffington Post asks Kurzweil his thoughts on love and spirituality, he responds:

"There are several different perspectives on love. One is that it is a high level concept of which our cortex is capable of. We have different pattern recognizers some will recognize that a capital A has a cross bar, those are kind of at the bottom of the conceptual hierarchy, and then some are recognize things like "oh that's funny," or "that's ironic," "she's pretty," or "I love that person" -- these are very high level concepts. When I say that computers will match human intelligence, I'm really talking about those concepts, because if you're talking about the logical ones, computers are already better than we are. But those are actually the cutting edge of human intelligence. And you can view it in an evolutionary perspective that there is actually value in us connecting to each other, because it leads to better protection of the species if we actually create communities based on love and act as a super organism as opposed to every individual for him or herself, which is actually not a good strategy overall.

So there is an evolutionary reason why love evolved. But I think that as we enhance ourselves through our technology and ultimately actually merge with it, we are going to be more capable of being more loving and capable of representing what that means. That is the epitome of what evolution is trying to achieve and in that regard, I consider evolution to be a spiritual process because it moves towards greater levels of intelligence, beauty, creativity, knowledge and love. All of which are attributes that god has been called, without limit, god is infinite in these qualities, an evolutionary process even the singularity which will be an explosion of these attributes that doesn't reach infinite levels, although there is some debate about that, that's my position anyway.

My view is it just grows at a doubling exponential rate, so it gets to pretty fantastic levels, but it doesn't become infinite, it may appear that way, from sort of our lowly vantage point. But it's moving in that direction, so we are becoming more godlike. So it is a spiritual process."

Now, personally what I found profoundly disturbing is anyone who is against technological progress, scientific research or that says we shouldn't extend human life? Really, I mean life expectancy according to most scientists and records that date back average the life expectancy to be about 47. Today it has doubled, however I predict in the next 20 years, we will go past what Dr. Aubrey De Gray calls the longevity escape velocity. What this means is that for every year you live another year, you will add another year to your life. Thus, you will be able to live forever.

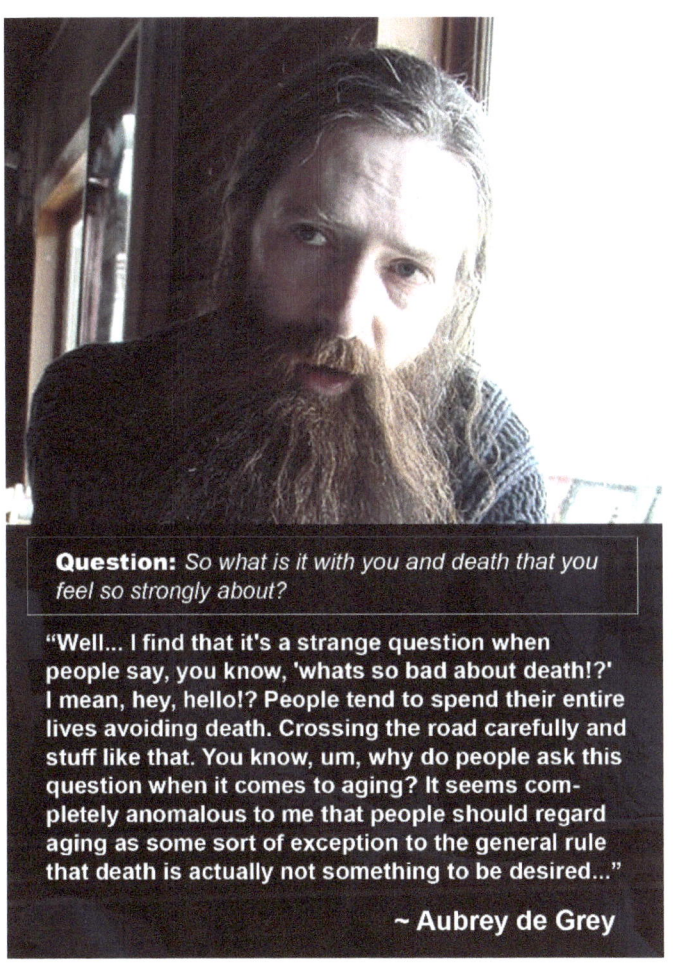

Question: *So what is it with you and death that you feel so strongly about?*

"Well... I find that it's a strange question when people say, you know, 'whats so bad about death!?' I mean, hey, hello!? People tend to spend their entire lives avoiding death. Crossing the road carefully and stuff like that. You know, um, why do people ask this question when it comes to aging? It seems completely anomalous to me that people should regard aging as some sort of exception to the general rule that death is actually not something to be desired..."

~ Aubrey de Grey

So where are we in terms of living forever? Well maybe one person who may be part of the solution appears to be my friend Dmitri Itskov and his 2045 project. I'm sure it's a social movement inspired by the movie Surrogates, Avatar and Ray Kurzweil's predictions, from speaking on the internet Mr. Itskov appears sincere, however we will have to see how this this goes and how true it is in its ambitions for being an ideology to being part of the solution to living forever and making life better.

First of my friend Dmitri is a Russian Mogul in the media industry with some online news websites and so forth, so yes he has some established "Klout". According to his own website *Russia 2045*, was founded by Dmitry Itskov, president of New Media Stars, in February of 2011 with the participation of leading Russian specialists in the fields of neural interfaces, artificial organs, and systems. Its major goal is creation of an artificial body, while the intermediate goal is to create various artificial organs, and life expectancy extension In fact he's even got the endorsement of my hero Steven Segal, yes the actor and 7 time Aikido champion.

Steven Segal, writes in his open letter to Vladmir Putin, about using and developing new technologies hopes to significantly improve the quality of life and make humans immortal by the year 2045. He writes:

"Not so long ago, I learned from my Russian friends about "Russia 2045" social movement, and I decided to join it" says Seagal, who has Russian roots and "considers himself Russian". "... I am extremely proud this technology has been discovered in Russia and that it has been propagated in Russia, with some of the greatest Russian doctors and scientists in the world. In bringing this technology into the forefront of the world, surely we as a human race will make incredible leaps and bounds ensuring a new and better quality of life,"

"I am appealing to you, hoping that we may have the opportunity for a mutually beneficial enterprise making the world a better place," Steven says. "It seems as through now you are placing more emphasis on life expectancy and life extension issues. I can see that you are actively working hard on coming up with solutions that lead Russia into the future confidently."

Steven Seagal has met Vladimir Putin in Russia and considers him "a prominent world leader" and hopes for his response. "Russia 2045" movement has been founded in 2011 in Moscow by the ground of prominent scientists. They intend to combine all existing groundbreaking medical technologies and develop new ones in order to produce artificial human body and, eventually, allow humans reach immortality.

Straight from their website and my friend Mr. Dmitri Itskov, the actual "2045" Initiative was founded by Dmitry Itskov in February 2011 in partnership with leading Russian scientists.

They maintain that their primary objectives are: the creation of a new vision of human development that meets global challenges humanity faces today, realization of the possibility of a radical extension of human life by means of cybernetic technology, as well as the formation of a new culture associated with these technologies.

The "2045" team is working towards creating an international research center where leading scientists will be engaged in research and development in the fields of anthropomorphic robotics, living systems modeling and brain and consciousness modeling with the goal of transferring one's individual consciousness to an artificial carrier and achieving cybernetic immortality.

An annual congress "The Global Future 2045" is organized by the Initiative to give platform for discussing mankind's evolutionary strategy based on technologies of cybernetic immortality as well as the possible impact of such technologies on global society, politics and economies of the future.

In fact from their website I've also listed a timeline they propose of this social movement of allowing scientists to meet each other and even fund each other's projects with the primary purpose of creating cybernetic Avatars and mind uploading.

I've personally chatted with Dmitri Itskov and he seems sincere as of now in promoting cybernetic technologies and mind uploading and research into Brain machine interfaces and promoting new breakthroughs and research to create a social movement where scientists can come together and also organize conferences to share their discoveries and also raise funding. I really believe he will succeed.

In fact on their website 2045, Itskov illustrates even a timeline of his proposed plan as listed below:

2015-2020

The emergence and widespread use of affordable android "avatars" controlled by a "brain-computer" interface. Coupled with related technologies "avatars' will give people a number of new features: ability to work in dangerous environments, perform rescue operations, travel in extreme situations etc.
Avatar components will be used in medicine for the rehabilitation of fully or partially disabled patients giving them prosthetic limbs or recover lost senses.

2020-2025

Creation of an autonomous life-support system for the human brain linked to a robot, 'avatar', will save people whose body is completely worn out or irreversibly damaged. Any patient with an intact brain will be able to return to a fully functioning bodily life. Such technologies will greatly enlarge the possibility of hybrid bio-electronic devices, thus creating a new IT revolution and will make all kinds of superimpositions of electronic and biological systems possible.

2030-2035

Creation of a computer model of the brain and human consciousness with the subsequent development of means to transfer individual consciousness onto an artificial carrier. This development will profoundly change the world, it will not only give everyone the possibility of cybernetic immortality but will also create a friendly artificial intelligence, expand human capabilities and provide opportunities for ordinary people to restore or modify their own brain multiple times. The final result at this stage can be a real revolution in the understanding of human nature that will completely change the human and technical prospects for humanity.

2045

This is the time when substance-independent minds will receive new bodies with capacities far exceeding those of ordinary humans. A new era for humanity will arrive! Changes will occur in all spheres of human activity – energy generation, transportation, politics, medicine, psychology, sciences, and so on.

Today it is hard to imagine a future when bodies consisting of nanorobots will become affordable and capable of taking any form. It is also hard to imagine body holograms featuring controlled matter. One thing is clear however: humanity, for the first time in its history, will make a fully managed evolutionary transition and eventually become a new species. Moreover, prerequisites for a large-scale expansion into outer space will be created as well.

Below is his yearly plan

- Creation of the first version of a human-like robot with a remote control and a telepresence effect

- Launch of a laboratory for research and development of Micro-Neuro-Interfaces

- Completion of the preliminary stage of designing the human brain life support system

- The preliminary stage of the human brain reverse engineering (Rebrain)

- Create infrastructure for a charitable foundation in Russia and the United States

- creation of a social online network *Immortal.me*

- organize and carry out Congress' Global Future 2045 "-2013 in New York City

- Organize and carry out '2045' master classes, workshops and courses on neuroscience, robotics, artificial intelligence theory, etc.

- Competition for student theses in the context of 'Avatar' project at the leading universities in Russia

- Creation of an analytical center for monitoring and analysis of advanced projects in the areas of NBIC convergence

- Creation of an international ranking of countries and projects, either directly or indirectly related to the development of cybernetic immortality technologies

- Filming a documentary about "2045" strategy.

Key elements of the project in the future

- International social movement
- social network immortal.me
- charitable foundation "Global Future 2045" (Foundation 2045)
- scientific research centre "Immortality"
- business incubator
- University of "Immortality"
- annual award for contribution to the realization of the project of "Immortality".

2045 AVATAR PROJECT MILESTONES
STRATEGIC SOCIAL INITIATIVE
Avatar D 2040 - 2045
A hologram-like avatar
Avatar C 2030 - 2035
An Avatar with an artificial brain in which a human personality is transfered at the end of one's life
Avatar B 2020 - 2025
An Avatar in which a human brain is transplanted at the end of one's life
Avatar A 2015 - 2020
A robotic copy of a human body remotely controlled via BCI
2045.COM

СОЗНАНИЕ
НАКАНУНЕ
РЕВОЛЮЦИИ

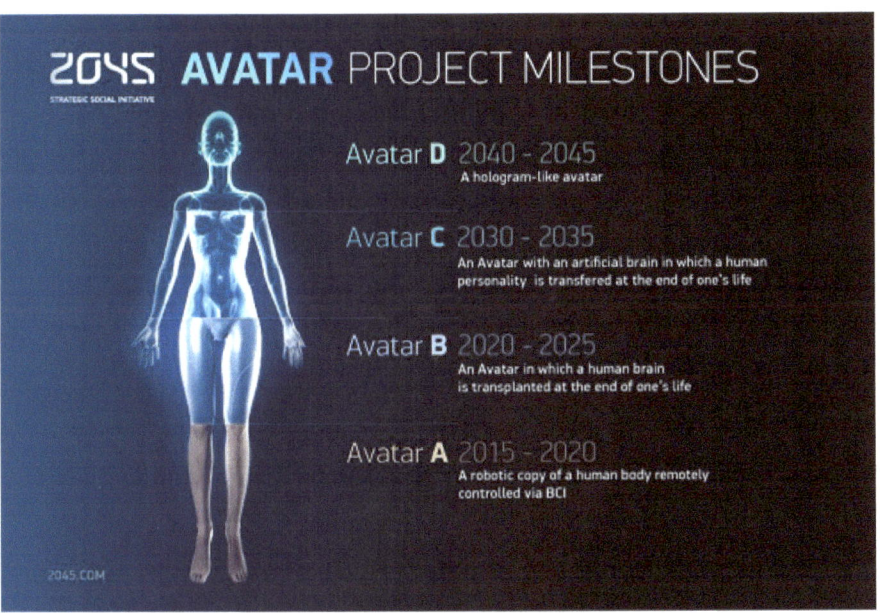

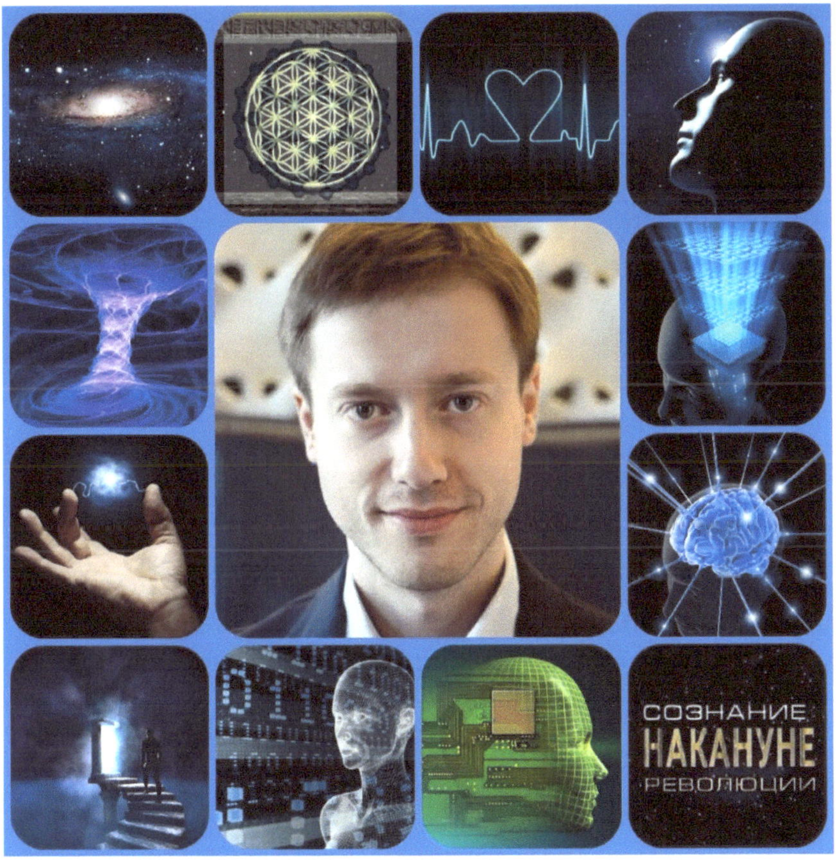

So what technologies will enable us to get to the singularity and what sort of radical breakthroughs should you keep your eyeball on?

Well first of is the personal genomics revolution. Gene sequencing was simply an affair that was too expensive for most people when the first one took forever to get sequenced over a decade and cost 3 billion dollars. Yes that's with a billion with a B. Can you imagine in the last decade that cost has fallen exponentially to 1000 dollars and in the next 3 years it will be 100 dollars or even less and happen much faster!

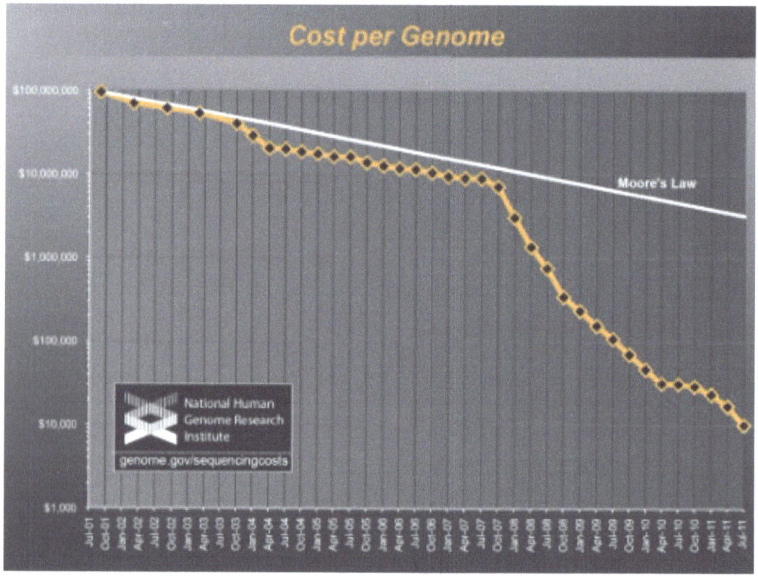

Source: http://g.foolcdn.com/img/editorial/SequenceCosts490.png

Basically, gene Sequencing involves determining the order of the bases — the chemical units of DNA that are represented by the letters A, C, G and T — in a gene or on a chromosome. The sequence helps determine the traits of an organism.

Some modern machines usually detect the bases by attaching fluorescent dyes to them and using a camera. But a new company called Ion Torrent that builds a sequencing machine uses a silicon chip that can detect the hydrogen ions given off when a new base is added to a strand of DNA. In essence they are using the collective one trillion dollar investment leveraged in the semiconductor industry to help sequence genomes.

Sequencing a sample on their machine takes only an hour or two, compared with days on many larger machines.

But preparing the sample can take more than a day, and the equipment to do that costs almost as much as the sequencer itself. Moreover, the Ion Torrent machine sequences only about 10 million bases per run, compared with billions for some other machines. The cost per unit of DNA sequenced is extraordinarily high. It costs $500 per run, in part because a new $250 chip is required each time.

Moreover, the new Ion Torrent machine can run a full human genome at 1000 dollars. Moreover, this cost is only expected to come down at a rate equal to or faster than Moore's law which means we will soon have a 100 dollar genome within a couple years. At that point everyone will have their genome sequenced possibly a few times and the big data of 7 billion genomes of 7 billion people will I predict prove to be a valuable resource. I believe that this will call for an age of supercomputers and computing power never previously required. Genomics and then also how genes are expressed and translated into proteins and the entire process is something that's of great value. It's this knowledge that will enable us to understand the cause of disease in humans. Moreover, scientists like Michio Kaku soon believe that a massive subtraction problem between the genomes of older and young people will enable us to figure out the genes that contain aging and we will use nanotechnology in 15 years I predict to silence genes off and turn on and express certain genes to control our biological process at the genomic level.

"The design of the human brain, while not simple, is nonetheless a billion times simpler than it appears, due to massive redundancy. Biological systems such as the brain are probabilistic fractals, which give them their messy, unpredictable quality. The design of the human brain is in the genome, and I show that there are only about 50 million bytes of design information (after lossless compression) in the genome (including the epigenetic information in the reproductive machinery), which is a level of complexity we can handle. And indeed, we are making exponential gains in modeling and simulating extensive regions of the human brain, including the cerebral cortex." — Ray Kurzweil

So can we replicate the human mind in all its complexity and if so what are the implications and possibility. Perhaps, I shall start with a thought experiment. What constitutes a human? I would say that we are a pattern of information, it's the way our brain is wired and the way it operates that gives rise to the emergent properties of humanity. Thus, humans, our entire personality comes from the brain.

I believe like I've heard many futurists say all our thoughts are an emergent form of our connectome and neural connections in our brain. Our whole personality lives on this biological substrate because of its unique structures and chemistry. Our entire body has a turnover time of 6 months to years, by which I mean our cells are made up of new molecules every few months or so thus like a river which maintains the same flow patterns yet minute to minute is constituted of new water molecules our body gives life to our personality and knowledge. So if were to switch to a non-biological substrate our personality and our entire consciousness could live in this neo avatar body. Thus I would say in the future we can replicate human consciousness in a non-biological body.

ONE DAY IN 2048

Let's take a trip to the year 2048, Enter London, England.

Perched high above the beautiful and modern London skyline, Fredrick arises. He is one of the few that's remained human, however in our technologically modern society he's nothing but human, with the nanotechnology of the day, he's practically a cyborg, with nanobots circulating in his bloodstream keeping him healthy from the inside out, Fredrick lives in a world of true healthcare abundance. The computers in his body are a billion times more powerful than our mobile phones today, with Nano-engineered sensors, powered by his body constantly monitoring his blood glucose levels, acidity, oxygen and constantly monitoring for tumors, cancerous cells and instantly delivering high doses of toxic substances promoting apoptosis or cellular cell death to toxic cells.

Fredrick, awakes in the morning walking to the bathroom. As he walks up to his toilet, of the future, he pleasantly walks up to his toilet and it opens automatically detecting his presence. Upon completing urination and walking to his bathroom sink

Today as he walks out of his bedroom, his son Noddy asks father about these new power ranger Lego action figures. Frederick remembers back for a second about how he would desperately look at new magazines and would hearken for the day he could go to the store to get the latest toy. He hearkens back and remembers for a moment waiting in line only to figure out he didn't have enough money and some months they would be sold out for days! He mentions to Noddy, you know son we would have too...

Noddy interrupts, "Dad" he says I know life was hard, now can you make me these new lego figurines I want 16 of them for my collection and want to give my friend John 4 more for his new collection.

Frederick goes onto his OLED tablet and pushes print and wirelessly a perfect 3D print of his figures appear in 20 minutes.

Could this happen? In fact it already is, look at the number of 3D printers available and what they can print. I This guitar was printed totally using a 3D printer, other objects include, a violin, toys, and even objects with gears that rotate.

In fact today we can print an entire car on a 3D printer, today.

This car was printed totally on a 3D printer. It's called the The 'Urbee' was made using a special printer which built up layer upon layer of bodywork - almost as if the car was 'painted' into existence, except using layers of ultra-thin composite that are slowly 'fused' into a solid. But unlike most 'innovations' in cars, this one won't break down after 5 years - Urbee is built to last 30. The team has a goal to build the vehicle totally out of recycled materials soon too! So this is how the next billion cars shall be sold, incredibly cheap 3D cars with electric powertrains. In fact using new batteries with quick charging and ranges 10X that of today thanks to nanopore batteries cars like these would have 1000-1500 mile ranges on a charge. In fact those Tesla cars in the earlier chapters could soon be printed easily, in fact any car you design and custom build can be easily 3D printed by anyone with a 3D printer. This totally changes the personalization of cars. In fact you can 3D print many shells for cars like these and you can change whatever body you want for your car. Like an SUV for the weekend or sports car for the day.

Source: http://www.dailymail.co.uk/sciencetech/article-2041106/Urbee-The-worlds-printed-car-rolling-3D-printing-presses-.html#ixzz29VUJWYaB

Source: www.evworld.com

NOW BACK TO OUR DAY IN 2048

Frederick sighs with relief of his son's temporary joy. Frederick now walks to his living room but without looking Frederick falls out of a window tripping on a robotic vacuum cleaner that's a vintage from 2 decades ago and falls 3 floors down out his glass!
This would have been horrendous a few decades ago, but not in this day and age, immediately his vitals are uploaded to the cloud and his clothes embedded with Nano sensors calls the emergency services.

Quickly an autonomous ambulance arrives, a humanoid robot picks Frederick up, along with a Tricorder he's quickly diagnosed and transported to the hospital.

At the hospital, a full 3D imaging study is done in real time in a matter of minutes. Artificial blood is injected to keep his blood pressure up and his vitals. The AI doctor analyzes the data and an attending human physician realizes that Frederick requires a new spleen and had organ damage, quickly as he's stabilized, a new 3D printed organ is printed matched to his body and blood type. Using a surrogate avatar robot, the surgery takes place with a human guiding the surgeon in faraway India. The attending physician at the hospital observes to ensure quality control and as a backup for any problems on the fly. For now Frederick is saved and will quickly heal in a day and be discharged.

If any of this seems far-fetched I would maintain it's actually quite conservative and that technology will advance faster than most of my predictions in this book most likely based on the previous 40 years and an extrapolation of the exponential trends in technology, biological sciences and AI and robotics. In fact most of this book in the previous sections outlined the technologies that are the basis of the future and the trends you should be aware of if you want to survive the 21st century economy and neo humanity!

Maybe Arthur C. Clarke, the well-respected author says it best!

"When a distinguished but elderly scientist states that something is possible, he is almost certainly right. When he states that something is impossible, he is very probably wrong." — Arthur C. Clarke

Today we see most of the culture of the singularity is going mainstream. I mean turn on the television and you see people like Ray Kuzweil and Michio Kaku on television promoting the future of technology all the time. I've even seem them on Jimmy Kimmel and David Letterman and on the Colbert Report. Moreover, we are seeing more coverage of exponential technologies every day on the mainstream news finally. I mean look at the excitement every year of new smart phones or 3D printers or about those smarter robots being built by DARPA or those Google Glasses or Cars. It's every single technology that is the underlying result of exponential growth and the basis for the next generation of technologies.

**In fact many movies demonstrate well what our future will look like, I mean take, iRobot, AVATATAR, or my friend, Director Barry Ptolemy and his film of Ray Kurzweil called TRANSCENDANT MAN, or even the famous Bruce Willis movie Surrogates.
Thus, movies reveal the future is near.**

The rise of the bottom Billion(s)

What if I told you there are going to be over 2 billion new customers for some products and services and you had the tools right now to serve them. What if I told you that these same people right now were not the interest of the largest companies in the world, what if I told you had the power to help a billion people and give them the tools and services they needed and you could do it from anywhere in the world with tools you already have? You're probably really anxious now aren't you?

Most people always ask the question is the world getting better or worse? They ask, will the bottom 3 billion in nowhere near the shape of the middle class in America better off now than they were fifty years ago? Will the farmer or rickshaw driver in India be impacted by these technologies and better healthcare? In fact are we better off than we were 20 years ago? Well to best answer that question I would say read my good friend Dr. Peter Diamandis' book *Abundance in* that book Dr. Diamandis and co-author Steven Kotler examine the issues of the day and show empirical evidence that the world is actually getting better. We are living in times of abundance in many fields and humanity lives at a time when information and many vital resources are actually quite ubiquitous and cheap!

One good fact to know is that the next billion people in the world will get connected online and have access to all of human information with tools like this 35 dollar android tablet and many new android phones.

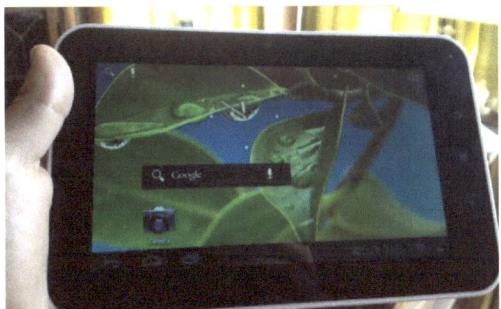

Source: http://techcrunch.com/2012/09/20/aakash-tablet/

If you're not focused about the bottom billion, the people that live on less than 1.50 cents per day. You should be, soon they will be online and consumers and users of the internet.

But first off what's the power of the internet? Well it means everything! Anyone with a sub 100 dollar smartphone or tablet like the Aakash 2 pictured above can access the world's best education on websites like EDX, Khan Academy, Udacity, and the billions of educational YouTube videos and even whole web at their disposal plus over 700k apps on Google play with one of these. They have roughly over 900 thousand dollars' worth of computation plus A.I. in a tablet with access to a knowledge graph and instant web page translation they can read any language on the web. If you're an app developer, you will have over a billion new customers that are coming online soon. A tablet or smartphone is worth everything to the poor because it gives them access to the same resources as millionaires in America.

There are so many ways to leverage the wealth of the developing world and an infinite amount of applications that for the near-term may help the next few billion people and at the same time make you wealthy. Here's one such application I've been working on, called the Circle Plus payments application, it will allow over a billion people who will have access to smartphones have access to mobile payments and be able to accept credit cards globally. You might be asking well these people aren't even on the internet yet? Well you see there lies the rub, in fact there are already over 5 billion mobile phones globally and in five years its estimated most of these phones will become smart phones without a doubt due to the doubling of smart phone adaption yearly and the 50% deflationary costs of mobile smart phones yearly.

For instance in countries like India there are roughly over 700 million feature phones, or "dumb phones". Meanwhile, the number of smartphones is roughly 100 million. However with sub 100 dollar phones, this adoption is doubling yearly, and will face growth rates faster than Moore's law in the coming years as phones drop to feature phone prices. Even more, what's astounding is that as the cost of mobile hardware, especially those running the android OS has fallen exponentially every year, those 700 million devices will soon be smartphones. Android devices are the most widely sold smartphones. There are already 500 million android devices globally in circulation. By 2015 it's expected, nations like India will have over 700 million smartphones. India is, after China, the second-biggest mobile market in the world, and it is growing fast, with some 6.5 million subscribers joining the ranks of mobile users in India in the month of April of 2012 alone, with around 80 percent of them on prepaid services, according to TechCrunch. Moreover, according to TechCrunch, there are about 280 million debit cards in circulation in the country, with an additional 18 million credit cards on top of that. But with a total population of over 1 billion, I believe this may be the largest opportunity is the payments arena globally. According to the website payment eye, mobile payments are expected to eclipse 1.3 trillion by 2017. Especially since almost no small merchants accept credit cards and rely solely on cash transactions!

Furthermore, currently the number of cards in the market is growing at a rate of 26 percent, with 50 million cards getting added last year. The fact that people are using them more for purchases, bill payments and other services is also having a knock-on effect to people getting more used to making other non-cash transactions around e-commerce. Already according to Business Insider, mobile internet usage has eclipsed desktop internet usage in India. This is the same trend that's holding true in all of Africa. According to TechCrunch, most Africans by 2015 will have smartphones, that's over another one billion consumers, we feel, by leading in India first, we can build our credentials to tackle mobile payments in India, all of Africa, Indonesia, Malaysia, and the Philippines.

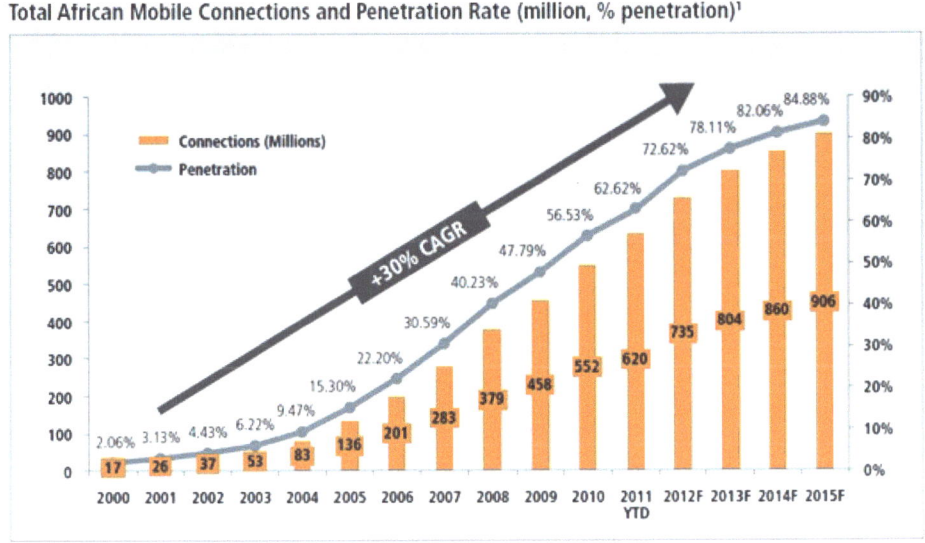

According to the info graphic below by

Source: http://www.roycefunds.com/news/global/2012/0125-africa-opportunities-challenges-growing-economy.asp

To be a successful entrepreneur in the technological age , you have to understand technological trends If you read the Singularity is near by Ray Kurzweil, you see his 31 year track record in technological trends, then you can understand timing is everything in making companies, most entrepreneurs fail because they're timing is wrong.
You have to leverage technological trends when they're at the knee of the exponential curve to reap the maximum benefits.

So what does this mean exactly? Well this means that anyone can become an entrepreneur and leverage these technologies as they emerge. For instance, as most of our purchases become intellectual properties instead of real purchases. For instance consider the fact that a large majority of our purchases are online nowadays, like movies, books, songs, and television shows. Many of us run websites that utilize advertising. One such company with an innovative idea poised to take advantage of this situation is a company called bitcoin.

One such company leveraging the revolution is Facebook is a prime example of a company starting a service at the right time.

Source: The Wall Street Journal

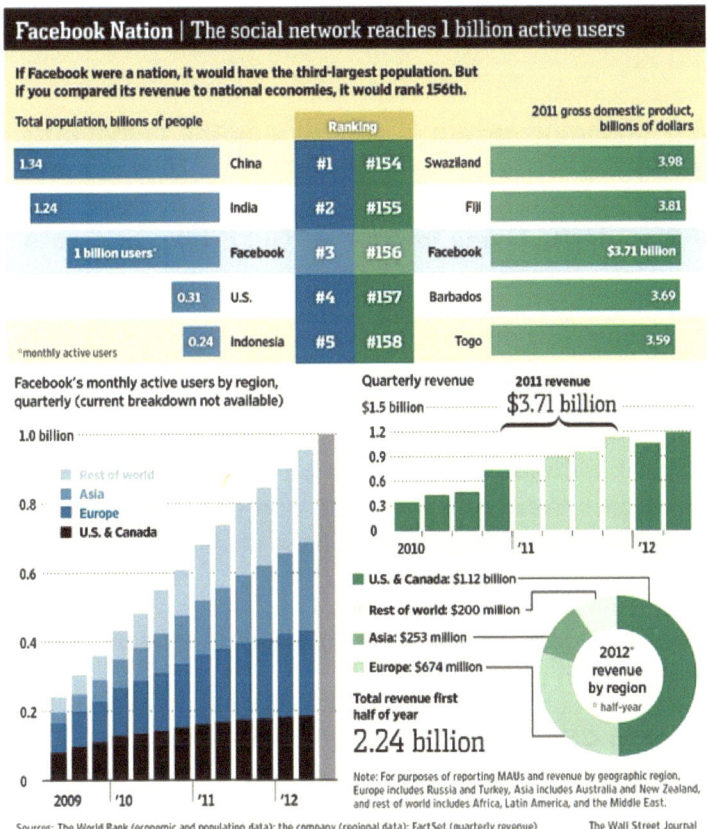

http://online.wsj.com/article/SB10000872396390443635404578036164027386112.html

The next question that arises is how an entrepreneur or company can stay relevant in this new age of exponential technology and how can CEO's and technology leaders stay ahead of the curve. Well first of reading this book should give some insight on how to think about the future of technology, but also a really good article by Megan Rose Dickey writes for the *Business Insider*, that one day, we'll inject tiny computers into our bodies like medicine or add them to our brains to make us smarter.

She maintains that in the meantime, tech is always getting faster, cheaper, and spreading to more markets and industries. And this creates a lot of opportunity for startups, until the day when we all turn into cyborgs. She continues,

"Because of this totally changing nature of society and the community business world, any company designed to succeed in the 20th century almost by definition has to fail in the 21st century," David S. Rose, Associate Founder of Singularity U and founder at Gust, tells Business Insider.

She asks next, so what does that mean for startups today?

In order to prepare for the singularity, Rose says, entrepreneurs need to figure out what technology will change and over how long, determine what effect that technology will have on a particular market, figure out what holes there will be to fill, and then actually build a business that will intercept that market hole when it comes around.

Amazon, Rose says, is the perfect example of a company that built a business with the singularity in mind.

She maintains that Amazon CEO Jeff Bezos foresaw a world where there was no longer a need for physical bookstores, so he decided to build one online. Once Bezos nailed down the distribution side of books, he had to start thinking about ways that competitors could kill his business. Given that the cost of storage, networks, and other digital technologies were dropping, Bezos realized the potential in digital books.

That was the birth of the Kindle.

Instead of waiting for a company like Apple to take him out, Bezos took himself out she laments.

"He deliberately shot himself in the foot because he knew that if he didn't do it, someone else would," Rose says.

Moreover, she says that "And someone eventually did. Apple announced in 2009 that it would be coming out with an iPad, and shortly after that, the tech industry proclaimed that the Kindle would die, but it didn't.

Even though Amazon doesn't release its exact number of Kindle sales, the company has continued to expand its Kindle lineup and announced in November that worldwide Kindle device sales over the holiday shopping weekend doubled.

Obviously, Amazon continues to face competition from the likes of Apple and Google. But Amazon is the perfect example of what a Singularity-focused business looks like, Rose says.

In short, here's how startups should prepare for the Singularity moving forward:
- Figure out where the ball will be a few years down the road.
- Determine how to hit that ball when it arrives.
- Figure out what could potentially take you out, and then take yourself out. "

Maybe the most memorable implications of all this new technology

may be what one of my friends Juan Carlos on Facebook says:

"If capitalism has successfully created highly intelligent, incredibly obedient, super controllable, and predictably tamed robots (human employees), there is no logical reason to believe technology won't be able to create even more intelligent, more obedient, more controllable, and more tamed robots based on silicon. Moreover, the emergence of Friendly AI will render capitalism obsolete. Imagine robots doing all the life's work of people in a tiny fraction of time: A Paradise on Earth! We won't need to work in order to survive. We will be on a permanent vacation to enjoy our lives. In the far future, money won't be necessary anymore: The production of goods will only cost matter/energy thanks to molecular manufacturing. And the production of services will only cost solar energy thanks to Strong AI. We will download everything from the Internet. For example: food, biological pets, virtual experiences, and more Strong AIs to solve all our problems. :)"

128

SAGA 3

Powering our human machine future

Do androids dream of electric power?

If all our civilization is predicated on technology and we're all going to be uploaded into supercomputers and live as surrogate robots, then how are we going to power this new human-machine civilization? That's the premise of this chapter.

Now, if there is one thing that I find really annoying, it would most definitely have to be when the Malthusians in power constantly apply linear thinking models to predict the future of energy. This pessimism pervades all the mainstream media. Turn the news on any given day or read many publications and all you will hear is that the price of oil is increasing or coal ash spills causing damage or energy shortages is all you seem to hear. Then naturally, the media goes on to mention how there's a water shortage or next about ever increasing CO_2 and other mainstream propaganda to control our lives. However, the sleeping elephant in the room is that the same technologies that are underlying semiconductor industry are prevalent in the energy sector too. However, looking at all the trends the only candidate at the time of the publication of this book that is subjugated to the law of accelerating returns appears to be solar energy.

On a side note what's remarkable is that in fact on our Earth all forms of energy are essentially solar power. I mean think of the biosphere all vegetation comes from some sort of photosynthesis and so forth. Coal is sequestered carbon that no one exactly knows how it forms but some say that burned decomposing wheat under pressure resembles coal, so scientists believe that coal is primarily organic plant matter crushed under high heat and pressure. Oil and natural gas have similar methods of production by our nature.

It appears that our future will be powered primarily by electricity and we are emerging as an electricity based economy. According to Ray Kurzweil, in his magnum Opus book *The Singularity is Near,* Kurzweil maintains that as a society we produce around 14 trillion watts of power today in the world. Out of this, 33% comes from oil, 25% from coal and 20% from gas, and 7% from nuclear fission reactors and 15% from biomass and only 1.5% from renewable sources of power. In fact oil subsidies in the form of defending the shipping lanes and militarism account for 500 billion US dollars in subsidies to oil according to prominent author and futurist Alex Lightman. What's even more frightening is the fact that according to Kurzweil, the total negative externalities of fossil fuel extraction account for over 2 trillion dollars in total damages in the forms of negative externalities to our environment.

    So now you hear the usual arguments against new technologies by the not so visionary bureaucrats and pseudo scientists in power. More often than not, you hear the arguments against clean energy or new technologies. In 2011 the large argument was solar is not economical or feasible. Just look at the failure of Solyndra was the popular one. If I had a dime for every time I heard that I'd be a billionaire. What's ironic though was that Solyndra failed because of the success of the solar industry. What? You're probably asking. Yep, that's correct, solar power has actually been growing exponentially for the last 31 years since 1981, for when we have the latest data. In fact the truth is, countries that rely on solar energy, free their market and use a decentralized grid will prosper the most, the future will require lots of power and we get over 10000 times more power than we use in a year from the sun in the form of photons and capturing that power is falling in cost by 50% every 18 months and has been for 31 years! It will be the one source of energy near-term in the next 5-10 years that will make energy truly too cheap to meter.

By the late 2020's and 2030, according to Kurzweil, the price-performance of computation will increase by a factor of ten to one-hundred million compared to the present day. However, all current calculations are based on linear models and don't account for the exponential increase of solar power, the only form of energy currently doubling every 2 years in capacity, price per watt, no matter how you plot it.

When I was a child, at just the age of 12, I noticed a very interesting trend that the amount of computational power per dollar was essentially doubling whether in RAM, hard disk space and so forth I became interested in predicting the future of technology, I made some videos and blogs as a child on these emerging trends and that they're remarkably easy to predict once you understand them. I read books by prominent futurists such as Theoretical physicist, Dr. Michio Kaku and then was blessed to discover a man who articulated much of my dreams in technology so eloquently in graphs and detailed studies, Ray Kurzweil and got a brief opportunity to speak to him at the Consumer electronics show on the power of exponential growth's implications and workings.

Why is this exponential growth so important you ask? Well it's because if you look at all forms of energy production and compare, you determine that of all the energy production in the world only one appears to follow the law of accelerating returns. One of my friends Raamez Naam, a computer scientist and author shared with me some graphs along with the DOE of energy that I was studying along with many technical publications I've read that actually calculated that solar power is essentially an information technology. One that doubles about every 18 months in price per performance, and then at the knee of the graph will really reach most of the world very cheaply. I see that our technological machine future won't be expensive, in fact due to the law of accelerating returns I figure we'll have cheap power for our lives and move into an era of abundance and that's just solar. By the

2030's energy I can say with almost 99% confidence for many people power will be too cheap to meter.

Here is a great paraphrased article from scientific American by Rameez Naam, a prominent author and former Microsoft employee that explains solar energy. According to the DOE, and Rameez Naam writes:

> The sun strikes every square meter of our planet with more than 1,360 watts of power. Half of that energy is absorbed by the atmosphere or reflected back into space. 700 watts of power, on average, reaches Earth's surface. Summed across the half of the Earth that the sun is shining on, that is 89 petawatts of power. By comparison, all of human civilization uses around 15 terrawatts of power, or one six-thousandth as much. In 14 and a half seconds, the sun provides as much energy to Earth as humanity uses in a day.
>
> The numbers are staggering and surprising. In 88 minutes, the sun provides 470 exajoules of energy, as much energy as humanity consumes in a year. In 112 hours – less than five days – it provides 36 zettajoules of energy – as much energy as is contained in all proven reserves of oil, coal, and natural gas on this planet.
>
> If humanity could capture one tenth of one percent of the solar energy striking the earth – one part in one thousand – we would

have access to six times as much energy as we consume in all forms today, with almost no greenhouse gas emissions. At the current rate of energy consumption increase – about 1 percent per year – we will not be using that much energy for another 180 years.

It's small wonder, then, that scientists and entrepreneurs alike are investing in solar energy technologies to capture some of the abundant power around us. Yet solar power is still a miniscule fraction of all power generation capacity on the planet. There is at most 30 gigawatts of solar generating capacity deployed today, or about 0.2 percent of all energy production. Up until now, while solar energy has been abundant, the systems to capture it have been expensive and inefficient.

That is changing. Over the last 30 years, researchers have watched as the price of capturing solar energy has dropped exponentially. There's now frequent talk of a "Moore's law" in solar energy. In computing, <u>Moore's law</u> dictates that the number of components that can be placed on a chip doubles every 18 months. More practically speaking, the amount of computing power you can buy for a dollar has roughly doubled every 18 months, for decades. That's the reason that the phone in your pocket has thousands of times as much memory and ten times as much processing power as a famed Cray 1 *supercomputer, while weighing ounces compared to the Cray's 10,000 lb bulk, fitting in your pocket rather than a large room, and costing tens or hundreds of dollars rather than tens of millions.*

If similar dynamics worked in solar power technology, then we would eventually have the solar equivalent of an iPhone – incredibly cheap, mass distributed energy technology that was many times more effective than the giant and centralized technologies it was born from.

So is there such a phenomenon? The National Renewable Energy Laboratory of the U.S. Department of Energy has watched solar photovoltaic price trends since 1980. They've seen the price per Watt of solar modules (not counting installation) drop from $22 dollars in 1980 down to under $3 today."

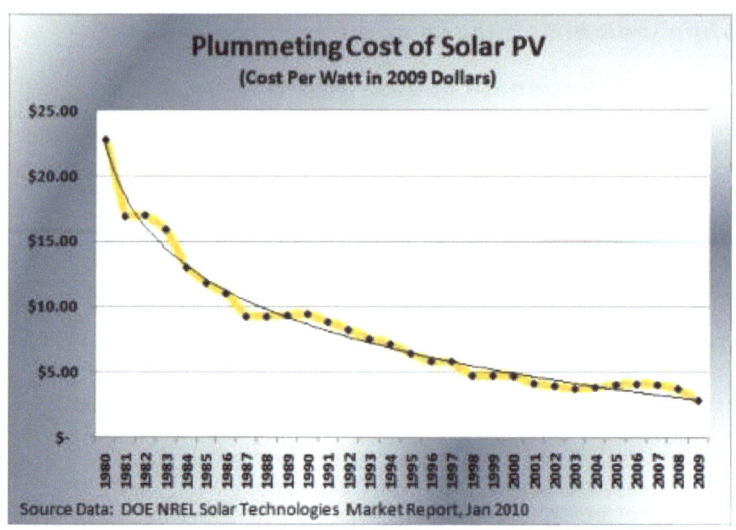

He later goes on to ask:

"Is this really an exponential curve? And is it continuing to drop at the same rate, or is it leveling off in recent years? To know if a process is exponential, we plot it on a log scale."

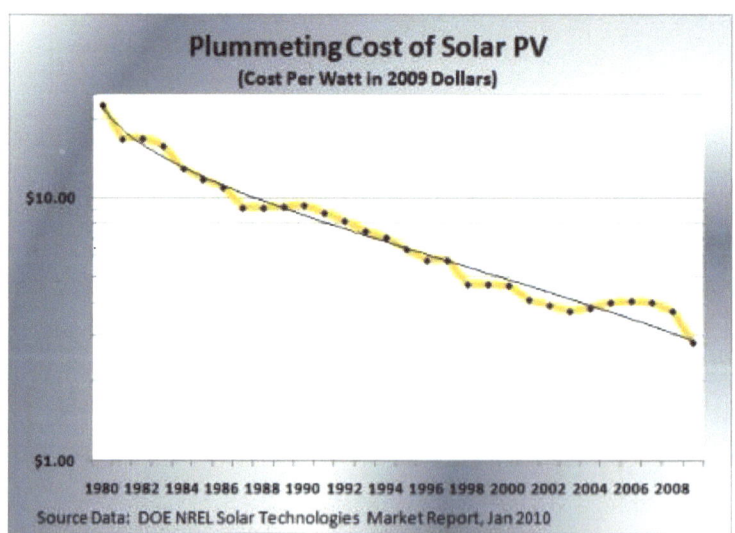

And indeed, he maintains that it follows a nearly straight line on a log scale. He goes on to mention:

"Some years the price changes more than others. Averaged over 30 years, the trend is for an annual 7 percent reduction in the dollars per watt of solar photovoltaic cells. While in the earlier part of this decade prices flattened for a few years, the sharp decline in 2009 made up for that and put the price reduction back on track. Data from 2010 (not included above) shows at least a 30 percent further price reduction, putting solar prices ahead of this trend.

If we look at this another way, in terms of the amount of power we can get for $100, we see a continual rise on a log scale."

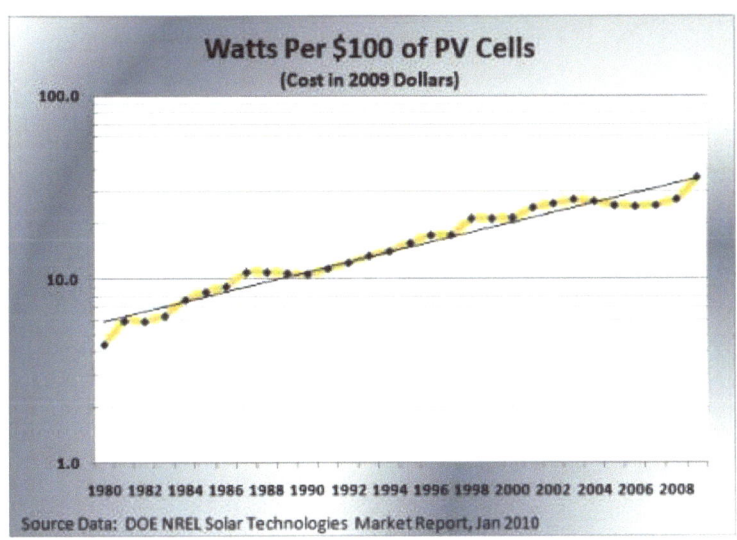

Next Naam asks what's driving these changes? He maintains:

"There are two factors. First, solar cell manufacturers are learning – much as computer chip manufacturers keep learning – how to reduce the cost to fabricate solar.

Second, the efficiency of solar cells – the fraction of the sun's energy that strikes them that they capture – is continually improving. In the lab, researchers have achieved solar efficiencies of as high as 41 percent, an unheard of efficiency 30 years ago. Inexpensive thin-film methods have achieved laboratory efficiencies as high as 20 percent, still twice as high as most of the solar systems in deployment today."

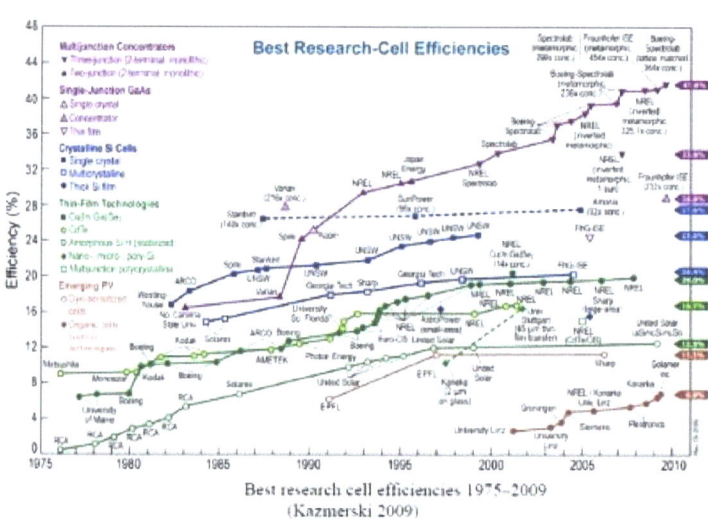

Best research cell efficiencies 1975–2009
(Kazmerski 2009)

What's astounding is when he mentions what these trends mean for the future? He says:

If the 7 percent decline in costs continues (and 2010 and 2011 both look likely to beat that number), then in 20 years the cost per watt of PV cells will be just over 50 cents.

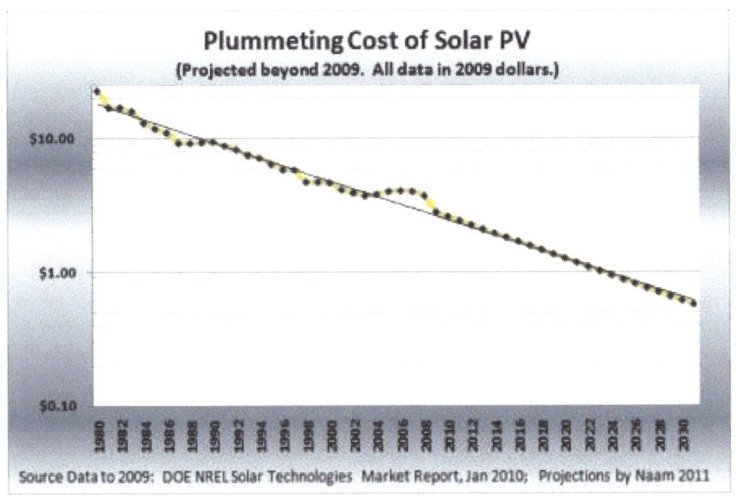

He continues to say:

Indications are that the projections above are actually too conservative. First Solar corporation has announced internal production costs (though not consumer prices) of 75 cents per watt, and expects to hit 50 cents per watt in production cost in 2016. If they hit their estimates, they'll be beating the trend above by a considerable margin.

What does the continual reduction in solar price per watt mean for electricity prices and carbon emissions? Historically, the cost of PV modules (what we've been using above) is about half the total installed cost of systems. The rest of the cost is installation. Fortunately, installation costs have also dropped at a similar pace to module costs. If we look at the price of electricity from solar systems in the U.S. and scale it for reductions in module cost, we get this:

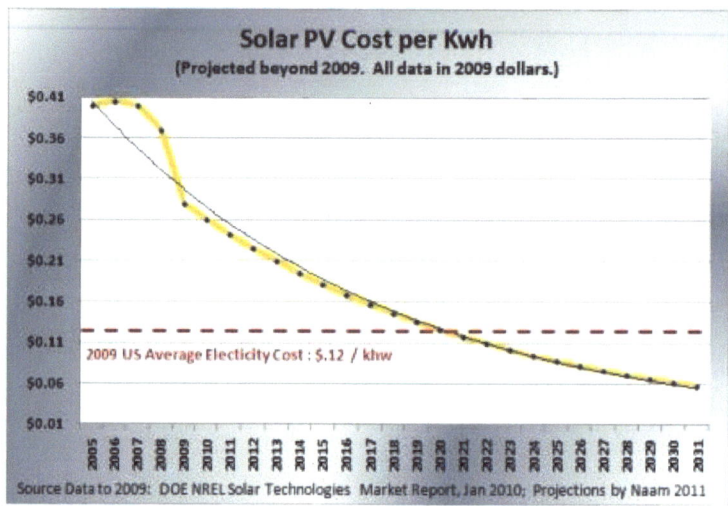

According to Naam:

The cost of solar, in the average location in the U.S., will cross the current average retail electricity price of 12 cents per kilowatt hour in around 2020, or 9 years from now. In fact, given that retail electricity prices are currently rising by a few percent per year, prices will probably cross earlier, around 2018 for the country as a whole, and as early as 2015 for the sunniest parts of America.

10 years later, in 2030, solar electricity is likely to cost half what coal electricity does today. Solar capacity is being built out at an exponential pace already. When the prices become so much more favorable than those of alternate energy sources, that pace will only accelerate.

We should always be careful of extrapolating trends out, of course. Natural processes have limits. Phenomena that look exponential eventually level off or become linear at a certain point. Yet physicists and engineers in the solar world are optimistic about their roadmaps for the coming decade. The cheapest solar modules, not yet on the market, have manufacturing costs under $1 per watt, making them contenders – when they reach the market – for breaking the 12 cents per Kwh mark.

The exponential trend in solar watts per dollar has been going on for at least 31 years now. If it continues for another 8-10, which looks extremely likely, we'll have a power source which is as cheap as coal for electricity, with virtually no carbon emissions. If it continues for 20 years, which is also well within the realm of scientific and technical possibility, then we'll have a green power source which is half the price of coal for electricity."

I was astonished to read this article from scientific American and was wondering about its potential implications. I figured the cost of solar power may fall even faster and continue to saturate much of our Earth and many local moons and planets like mars as it gets cheaper. This would make colonizing beyond our Earth feasible. This is just solar power and it will happen within this century.

In fact one such project underway is the solar roadways concept, which could power all of America and cover all our highways and it would pay for itself and last longer than current highways. Conceived by Julie and Scott Brusaw, and according to his website Scott is an electrical engineer (MSEE) with over 20 years of industry experience. This includes serving as the Director of Research and Development at a manufacturing facility in Ohio (developing their line of products for over 12 years), a voting member of NEMA (National Electrical Manufacturers Association), and developing several networked control systems from the ground up. Scott has multiple patents and his hardware and software have been sold internationally.

Here are the numbers straight from the concept's websites regarding its feasibility. Engineers love numbers. The author mentions that numbers generally bore people to death, but at times they are necessary for understanding. One of the biggest questions that has been asked is simply, "Can we really generate enough pollution-free electricity to power our businesses and homes?" The calculations below are presented to answer this very important question.

First, the "givens":
In the 48 contiguous states alone, pavements and other impervious surfaces cover 112,610 square kilometers-an area nearly the size of Ohio-according to research published in the 15 June 2004 issue of Eos, the newsletter of the American Geophysical Union. Continuing development adds another quarter of a million acres each year.

Let's do some conversions: 112,610 square kilometers equals 43443.54 square miles. The report was done six years ago, so that means an additional 1.5 million acres have been turned into impervious surfaces. That's an additional 2343.75 square miles, so all told, we have 45787.29 square miles of impervious surfaces. Let's make a conservative assumption that a full 1/3 of that number accounts for rooftops of homes and businesses, which we're not currently interested in. That leaves us with 28962.36 square miles of roads, parking lots, driveways, playgrounds, bike paths, sidewalks, etc., to work with.

If these impervious surfaces were replaced with Solar Road Panels, how much electricity could we produce?

In labs, solar cell efficiency has exceeded 42%, but you can't get your hands on these - at least not yet. For our calculations, we looked up commercially (and cost competitive) available solar panels.

Sunpower claims they will be shipping solar panels with 19.5% efficiency later this year (2010). They're already shipping E18 series panels with 18.5% efficiency, so let's go with what is currently available.

For our calculations, let's use the following assumptions:
. We use solar cells that have an 18.5% efficiency
. We average only 4 hours of peak daylight hours per day (4 x 365 = 1460 hours per year)

Sun power offers a 230 Watt solar panel rated at 18.5% efficiency. Its surface area is 13.4 square feet. If we covered the entire 28,962.36 square miles of impervious surfaces with solar collection panels, we'd get:

((28,962.36 mi²) x (5280 ft / mi)²) / (13.4 ft²/230W) = ((28,962.36 mi²) x (27,878,400 ft² / mi²)) / (13.4 ft²/230W;) = (807424257024 ft²) / (13.4 ft²/230W) = 13858774560860 Watts or over 13.8 Billion Kilowatts

If we average only 4 hours of peak daylight hours (1460 hours per year), this gives us: 13.8 Billion Kilowatts x 1460 hours = 20,233,810,858,855,600 Kilowatt-hours (or) 20,233 Billion Kilowatt-hours of electricity.

In 2009, we received a contract from the Federal Highway Administration to test some of our theories and to build a crude prototype Solar Road Panel. One of the tests that we conducted was "real world" solar collection.

When you install a solar panel, you have to take into account where you are installing it. The farther north you live, the more you have to angle your panel toward the equator (or more accurately, the sun above the equator). We did our testing in January and February in northern Idaho.

Here we have worst case scenario: our measurements were taken in the dead of winter (sun is at its lowest point of the year) an hour south of the Canadian border at latitude 48.19 degrees. The farthest northern point in the contiguous 48 states is 49.38 degrees near Lake of the Woods, Minnesota. That's 82miles farther north than our location. Conclusion: we would be hard pressed to find a worse time and place to conduct this experiment!

At our northern position (48.19 degrees North), the optimal solar gain angle for our solar panels is 72 degrees. Brownsville, Texans would want to angle their solar panels at 26 degrees. So our southern roads will naturally produce much more electricity than their northern counterparts, as solar intensity maps show.

Unfortunately, we can't angle roads or parking lots. Roads go up and down hills, have banks on curves (going both left and right), and have a typical three percent "crown" (on both sides) to allow stormwater runoff. It's a pretty safe assumption to figure that the national average angle of roads is zero degrees.

In fact, solar energy is the only candidate that firmly meets all the obligations needed for our electricity based future at a low cost. In fact, many questions get asked about solar, intermittency, is making the panels green, and many others like what happens when the sun doesn't sunshine?

I can answer all of these. For instance, of course what people don't understand is that just as solar panels are falling in cost due to the fact that they rely on the same advances and manufacturing capabilities of the semiconductor industry, they too can leverage existing factories and production lines using advanced robotics and cheap materials and in fact can even be made to be 100% recyclable after they undergo 20-30 year usage cycle.

Consider the fact that today solar power is already cheaper than the grid in 14 states as of this publication in America. How? Well in fact one such company called SolarCity installs solar panels and just charges you for the power at lower than grid rates in the 14 states they operate. In fact they've been so successful that in California already by this year there are already 120 stores that have solar panels on their roofs and Wal-Mart paid zero dollars for the panel's installation and they're even guaranteed to output power for 20 years! So Walmart paid zero dollars and pays less for the power each month than the grid costs, that's what I call a win-win-win.

In fact an even more ambitious project is taking solarcity actually just got a billion dollar financing totally from private banks to finance solar power on 120000 military homes across bases across America! Perfect win-win in the fact the military pays nothing for the panels or installation, all totally free courtesy of solarcity! Then they sell power at lower than grid rates! So again it's a win-win situation!

How? You're asking? Well Solarcity installs the panels and sells the electricity at 11 cents a kilowatt hour. In fact during peak times in California electricity rates can be up to 18 cents or more per kWh. Thus, According to Walmart's own press release they claim to have saved over a million dollars in electricity costs of doing this program. In fact to deal with times when the sun doesn't shine much or during night hours, Solar City is partnering with a company called Tesla motors, which makes electric cars to use their large battery packs to store power and save power for later usage.

In fact there is a company with such the solution for grid storage of solar power at a dirt cheap cost. It comes in the form of a new battery developed by Aquion Energy in Pittsburgh uses simple chemistry of a water-based electrolyte and abundant materials such as sodium and manganese and is expected to cost $300 for a kilowatt-hour of storage capacity, less than a third of what it would cost to use lithium-ion batteries. Third-party tests have shown that Aquion's battery can last for over 5,000 charge-discharge cycles and has an efficiency of over 85 percent. In fact the technology behind, Aquion's battery is that it uses an activated carbon anode and a sodium- and manganese-based cathode. A water-based electrolyte carries sodium ions between the two electrodes while charging and discharging. The principle is similar to lithium-ion, but sodium ions are more abundant and hence cheaper to use. Compared to solvent-based electrolytes, the aqueous electrolyte is also easier to work with and cheaper. Even better, the materials are nontoxic and the battery is 100 percent recyclable, Whitacre says. This is a technology that every company or warehouse or any building with a roof where Solarcity operates can buy power at lower than grid rates with zero dollars down and then store excess power with these cheap batteries. Best of in the 14 states SolarCity operates, they can do this today. I predict in the next 10 years almost all 50 states will have solar leasing as solar power drops by 50% about every 2 years to generate. In fact, I predict that people will just pay 2000-4000 dollars soon for 5KW systems to put on their house like roof tiles by the early 2020's which can power the average home and generate about 200 dollars' worth of electricity a month based on current average power rates. Since people buy appliances for two thousand bucks, why wouldn't you buy a 5 kW solar system with a payback of one year? By 2020 this will happen I predict. Furthermore, if you don't use the power you can always sell it back to most utilities earning a credit for your excess power. Already in Germany for instance many farmers have utilized their underperforming farm land and filled them with solar panels, the panels sell power into the grid and the farmers now receive free cash every month from their neighbors buying solar power they generate, talk about 21st century farming right? Solar farming that is.

Thus, I am going to make a bold prediction that by 2028 more than 80% of the power generated in America will be solar based due to the law of accelerating returns. Why am I confident when well today we get about 1.5% of our power from solar and renewables, but I expect solar power to double in capacity every 2 years, I predict we're on the knee of the exponential curve and by 2028 we will have a large share of our power be solar based.

Best of all as my friends say, renewables are a free lunch. We will never have any major energy crises as a result of a renewable energy sources drying up. That's because renewables are ultimately drawn from energy sources that will outlast the Earth itself.

What will all this cheap energy and power mean for oil? Well oil is one of the most useful substances as is plastic. In fact one turns out to be the other.

Well enter the 'Blest' machine—produced and distributed by Japanese Blest Company—is the brainchild of inventor Akinori Ito, and may resolve not only the world's diminishing fossil fuels problem, but also reduce plastic waste pollution. In fact most of our oceans are filled with plastic and a large majority of our garbage turns out to be plastic.

Just how is this going to be possible?
It's simple the Blest machine could be used to melt all sorts of plastic into oil. In fact if done on a large commercial scale using wasted heat or excess PV power, all the wasted plastic could be melted back into oil and be put into productive use.

Will cheap power mean cheaper transportation?

Now cheap power, naturally leads us to the car. I previously discussed the future of the car, predominantly, self-driving A.I. vehicles in the previous chapter. However, one emerging trend is the electric car. Now the electric car was invented well over a hundred years ago, and Thomas Edison predicted in fact we may all be having an electric car, in fact this now seems very true, and there are many emerging technologies that are making this possible.

In fact, there have been many improvements in AI and entertainment capabilities making improvements in cars, including voice guided navigation, self-driving cars such as the autonomous cars that Google has a whole fleet of and can drive through any roadway in America without driver input. Moreover great advances in cars have led to cars like the one I own in which almost every feature totally voice controlled. For instance, I can just say temperature 74 degrees or radio 107.1 and boom it changes. However, the powertrain is still stuck in the last century, still today we tend to use internal combustion engines to power our cars. However, this great stagnation isn't going to last too long. However, it appears that in terms of the powertrain, future cars will be electric, and not in a very long time frame either.

Meet Elon Musk, founder of PayPal, Chairman of Solarcity, CEO of SpaceX and Tesla motors, and claims that in 20 years almost half of all cars sold will be electric. Keep in mind that there are now 1 billion vehicles on the planet, with 2 to 2.5 billion likely by 2050. Also, the auto industry now builds 60 to 80 million vehicles a year, which is likely to rise to 100 million or more by 2020.

Globally, in 2012, fewer than 100,000 of those will be plug-ins. Musk's half-plug-ins-by-2032 guess is a worthy goal. Why, well first of all electric cars today have limited ranges, toxic batteries and are tiny and slow and no one wants one. However all this is changing nowadays.

Meet the Tesla Model S and Model X an SUV that can seat 7, do 0-60 in 4 seconds and has a 300 mile range according to Tesla. In fact I have some friends who've ordered them already. While speaking to George Blankenship during my interview at GigaOM Roadmap 2012, one of the big whig executives at Tesla, I learned about the demand and how much Tesla is going to grow both nationally and internationally. This year alone 13000 pre-orders and next year expanding production to 20000 units according to Elon Musk.

What's even more astounding is that Tesla sells a $35000 dollar Model, has a cybertruck coming out and a fleet of semi trucks which are currently in operation. If that becomes successful, every manufacturer will copy or buy tesla powertrains, and batteries and build electric cars, I predict Musk's guess of half of cars sold being either plug in hybrid or electric is not pie in the sky. In fact it only costs a few bucks to charge an electric car and in fact with solar power being ubiquitous soon powering your car will be free and pollution free. The best of both worlds. Moreover, can you imagine the fact that Tesla is building a supercharger network of charging stations where anyone with a Tesla can charge their car for free in 30 minutes coast to coast! Now that' what I call free long distance.

The Tesla Model S Sedan

-Courtesy Tesla Model X suv

This is what those superchargers look like from the Press event Tesla for demonstration!

Photo: courtesy of Tesla motors:

http://www.teslamotors.com/about/press/releases/tesla-motors-launches-revolutionary-supercharger-enabling-convenient-long-dista

MORE CHEAP POWER!

If we go further what I speculate is that two more forms of energy will pervade our life in the near future. One of these is quantum vibrations, as our civilization faces a post biological transformation, our energy sources will fall until we hit the laws of physics as a baseline and use very little power for a lot of computation. As computation surrounds us.

However the one technology that will make humanity a type 1 civilization and that will allow us to spread humanity, which according to me is our destiny as the human species, is that of nuclear fusion. Nuclear fusion has always been a technology that's stuck in the realm of science fiction. The common joke among scientists I've seen talk about fusion is that fusion is 10 years away and always will be. However, fusion happens all around us. The sun is the largest fusion reactor in our side of the Milky Way galaxy. I truly believe that in the next 15-20 years major breakthroughs will enable the start of at least a few fusion plants for trial that actually output more power than we need. If we succeed, the human civilization can be sustained anywhere in the universe with such an infinite and renewable power source with almost now drawbacks or nuclear waste. That's a true 21st century solution to energy.

SAGA 4

LET'S GO GALACTIC!

The most important transformation we will make over the next century that we all will be a part of is becoming a type 1 civilization. I believe Peter Garretson, of the USAF (United States Air Force) says it best when he maintains that humanity needs a billion year plan. In fact I totally agree.

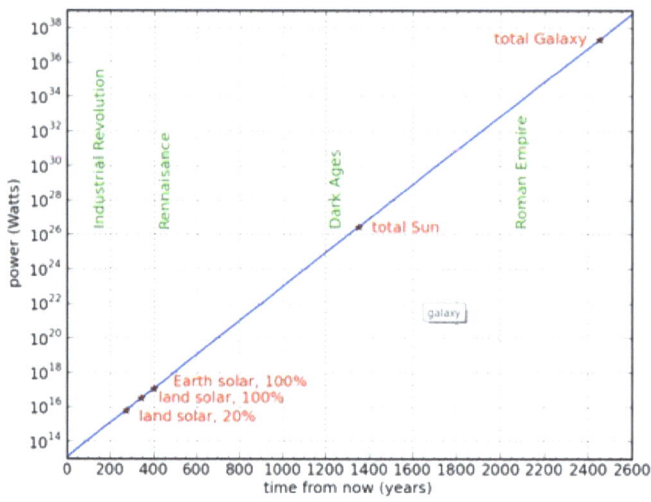

Source: http://www.kurzweilai.net/what-our-civilization-needs-is-a-billion-year-plan

As I've said earlier in the book we get 10000x more power from the sun than we use, however, we will soon within the next 50 years use space based solar power which is very feasible but expensive for now. But even covering all the empty warehouse roof space and 1% of the deserts with solar power provide more power than we will need for the next 100 years. Why will we need this sort of power? Well many of us will upload our personalities and consciousness to computers as we replace our biological neurons with nanobot based synthetic neurons and gain the capacity to be able to back up our mind information into the cloud. At that point we will become an electricity dependant civilization. We will need large amounts of power to transform the moon and asteroids and space minerals and convert much of the Earth to intelligent computation. David Brin, an advisor to the Lifeboat foundation maintains that science fiction author Vernor Vinge has been touted as a chief popularizer of this notion, though it has been around, in many forms, for generations. More recently, Ray Kurzweil's book *The Singularity is Near* argues that our scientific competence and technologically-empowered creativity will soon skyrocket, propelling humanity into an entirely new age.

David Brin, describes the Singularity best when he says, "Call it a modern, high-tech version of Teilhard De Chardin's *noosphere apotheosis*, an approaching time when humanity may move, dramatically and decisively, to a higher state of awareness or being. Only, instead of achieving this transcendence through meditation, good works or nobility of spirit, the idea this time is that we may use an accelerating cycle of education, creativity and computer-mediated knowledge to achieve intelligent mastery over both the environment and our own primitive drives."

What are some benefits?

- When we start using nanotechnology to repair bodies at the cellular level?
- When catching up on the latest research is a mere matter of *desiring* information, whereupon autonomous software agents deliver it to you, as quickly and easily as your arm now moves wherever you wish it to?

- When on-demand production becomes so trivial that wealth and poverty become almost meaningless terms?

- When the virtual reality experience — say visiting a faraway planet — gets hard to distinguish from the real thing?

- When each of us can have as many "servants" — either robotic or software-based — as we like, as loyal as your own right hand?

- When augmented human intelligence will soar and — trading insights with one another at light speed — helping us attain entirely new levels of thought?

These are some of the benefits of uploading into the cloud. We no longer need food, money or other resources.

Notes:

A lot of you are probably getting worried about the implications of the singularity, to show that technology has both problems and peril. Probably the best way to describe the ethics and the fact technology is a double-edged sword is from a paper I wrote on the implications and ethics concerns of the coming technological singularity.

Promise and the Ethics of the coming Singularity

In the future as Moore's law progresses, which basically means that computer power doubles and price halves every one to two years, the human civilization will basically reach technological Singularity, which primarily means that computing power will create artificial intelligence that is as a capable and intelligent as the Human Brain. Moreover, as artificial intelligence progresses society will have developed computers that are "awake" and superhumanly intelligent. Furthermore, technology in this field will bring great change to the human civilization. For instance, for the first time in history we will have human computer interfaces that are so intimate that users may be considered "superhumanly intelligent." Out of all this technology, however, the one invention that will "rewrite" all rules of humanity will be the invention or concept of total Mind Uploading. This is the hypothetical process of scanning and mapping a biological brain in detail and copying its state into a computer system or another computational device. The computer runs a simulation model so faithful to the original that it will behave in essentially the same way as the original brain, or for all practical purposes, indistinguishably. First of all to ensure this level of intelligence a machine must pass a Turing test, which means a machine must have the same level of intelligence and reasoning as a human. If the intelligent machines pass the test then the human civilization will have reached a great paradigm shift. However, with the invention of such a technology we must as a civilization question the moral ethics of such a technology and evaluate the pros and the cons of such a profound change it would have upon humanity. Many of the benefits that it would bring include human immortality, elimination of all biological diseases, as we will live as a avatars of ourselves on computers and be able to make endless copies of ourselves, "backup" all our knowledge and intelligence and even be in multiple places at once, creating endless possibilities to do an innumerable amount of tasks at the same thing. However with the benefits such a technology of total mind uploading will bring, it is imperative to consider the number repercussions of such a technology on the human civilization and the ethical implications this will bring for humans. For instance, some questions to Implore are will we still be considered human if we are living in machines? Moreover, in the age of total mind uploading and memory enhancements, will this technology give people that choose to adapt and not be human, in the traditional sense, an unfair advantage in society? Thus, in this examination, we will examine how total mind uploading would be possible, then evaluate the ethical implications and then examine the pros and cons of such a technology in the future. Hence, in this analysis, let us examine the potential benefits and harmful repercussions of such a technology on humanity.

To begin, let us first implore how such a technology would be possible and work in the future, before we begin to examine the potential implications upon humanity of total mind uploading. When tracking technology trends, well known Futurist, inventor and founder of Kurzweil Technologies, Ray Kurzweil says, "Since 1890, computing power has become a trillion times more powerful, and a billion times more powerful in the last 25 years. The computer I used at the beginning of college had 10 times less power and half the storage space of the one I used as a senior. With the human brain able to hold only about 2 petabytes of information, a single computer will equal the storage capacity and speed of the human brain by around 2029 (Fox)." Moreover, when tracking technology trends it can easily be seen price per performance ratio of computers, approximately every year price halves and performance doubles (Kurzweil). This paradigm of Moore's law also applies to temporal resolution of Brain-scanning, nanotechnology, and supercomputer performance (Kurzweil). Dharmendra Modha, manager of cognitive computing for IBM Research says a simulation of a human cortex could come within the next decade if Moore's Law holds (Robertson). Moore's Law is the rule of thumb that the number of transistors on a computer chip tends to double every two years. Thus, from this we can easily deduce that a "digital Brain" is near and that we will have total mind uploading in the near future.

Moreover, to add to the evidence of how this would be possible and that it would not necessarily be unethical, Kurzweil maintains that first "this will not be an alien invasion of intelligent machines. It will be an expression of our own civilization, as we have always used our technology to extend our physical and mental reach. For instance, we began this quest with our supercomputers and then mobile phones. We will merge with this technology by sending intelligent nanobots (blood-cell-sized computerized robots) into our brains through the capillaries to intimately interact with our biological neurons. If this scenario sounds very futuristic, I would point out that we already have blood-cell-sized devices that are performing sophisticated therapeutic functions in animals, such as curing Type I diabetes and identifying and destroying cancer cells. We already have a pea-sized device approved for human use that can be placed in patients' brains to replace the biological neurons destroyed by Parkinson's disease, the latest generation of which allows you to download new software to your neural implant from outside the patient (Kurzweil)." Thus, from this we can easily deduce some of the inherent benefits of total mind uploading. These will include always having a virtual backup of all our thoughts and knowledge. For instance, today we would not dream of not backing up or saving on a secure server all our personal data on our computers. Similarly in the future it will be imperative to preserve all our thoughts and knowledge in the virtual world. Moreover, consider the potential benefits that having a digital backup of our thoughts would have on our lives. No longer would an athlete or businessman who was injured be deemed useless after an injury. Simply could they restore their thoughts to a surrogate body or robot and carry out their tasks through that avatar of themselves. This in turn would exponentially increase GDP and economic productivity as sick days and injury absences would literally become a thing of the past. The reason being no other than the fact that injuries, sick days, loosing productivity would no longer be an excuse. To put this simply if you get hurt transfer all your mental capabilities to a surrogate body and continue working. Hence, it is possible to see that total mind uploading would eliminate almost all of the problems facing workers in the workplace.

References

SAGA 1

BUILDING A KNOWLEDGE BASED ECONOMY

Cover image

http://foftw.org/2010/08/23/web-3-0-the-future-of-internet/future-city-5-web/

http://theeconomiccollapseblog.com/archives/10-things-that-every-american-should-know-about-the-federal-reserve

Enriquez, Juan. *As the Future Catches You: How Genomics & Other Forces Are Changing Your Life, Work, Health & Wealth*. New York: Crown Business, 2001. Print.

http://www.techdimwit.com/4/post/2009/9/this-is-the-transcript-of-juan-enriquez-speech-on-technology-and-genomics-its-extremely-interesting-for-all-you-techdimwits-out-there.html

Enriquez, Juan. *Juan Enriquez: The life-code that will reshape the future*. 2003. Graphic. TED TalkWeb. 9 Sep 2012.

http://www.nationalreview.com/articles/278758/end-future-peter-thiel

http://hplusmagazine.com/2010/02/24/next-global-superpower-korea/

http://www.philly.com/philly/entertainment/20120516_Sacha_Baron_Cohen_rsquo_s_lsquo_The_Dictator_rsquo__hilariously_weighs_democracy__autocracy.html#ixzz280JmDJb6 Watch sports videos you won't find anywhere else

http://www.economist.com/node/21540395

SAGA 2

Kurzweil, Raymond (2005). *The Singularity is Near*. Penguin Books. ISBN 0-14-303788-9.

http://www.smartplanet.com/blog/bulletin/automation-will-soon-touch-every-job-on-the-planet-prediction/968

Singh, Avinash. How to Control a Prosthesis With Your Mind. Web. 8 Oct. 2012

http://mil-labs.blogspot.in/2012/10/how-to-control-prosthesis-with-your-mind.html

Beck, Glenn, Cowards: What Politicians, Radicals, and the Media Refuse to Say. New York: Threshold Editions/Mercury Radio Arts, 2012

Luke Muehlhauser and Anna Salamon, "Intelligence Explosion: Evidence and Import"

Daniel Dewey, "Learning What to Value"

Eliezer Yudkowsky, "Artificial Intelligence as a Positive and a Negative Factor in Global Risk"

Luke Muehlhauser and Louie Helm, "The Singularity and Machine Ethics"

Luke Muehlhauser, "So You Want to Save the World"

Michael Anissimov, "The Benefits of a Successful Singularity"

http://www.marshallbrain.com/robotic-nation.htm

http://www.mdm.com/economy/2012/02/22/automation-not-outsourcing-biggest-drain-on-manufacturing-employment/PARAMS/post/28358

1. **R. Buckminster Fuller,** *Nine Chains to the Moon*, **Southern Illinois University Press**[1938] 1963 pp. 276–79.
2. **R. Buckminster Fuller,** *Synergetics (Fuller)*,http://www.rwgrayprojects.com/synergetics/s04/figs/f1903.html
3. Moravec, Hans (1998). **"When will computer hardware match the human brain?"**.*Journal of Evolution and Technology* 1. Archived from the original on 15 June 2006. Retrieved 2006-06-23.
4. Moravec, Hans (June 1993). **"The Age of Robots"**. Archived from the original on 15 June 2006. Retrieved 2006-06-23.
5. Moravec, Hans (April 2004). **"Robot Predictions Evolution"**. Archived from the original on 16 June 2006. Retrieved 2006-06-23.
6. Ray Kurzweil, *The Age of Spiritual Machines*, Viking, 1999, p. 30 and p. 32
7. *The Law of Accelerating Returns*. Ray Kurzweil, March 7, 2001.
8. http://accelerating.org/articles/huebnerinnovation.html
9. e.g., Johansen, A., and D. Sornette. 2001. Finite-time Singularity in the Dynamics of the World Population and Economic Indices. *Physica A* 294(3–4): 465–502

^ e.g., Korotayev A., Malkov A., Khaltourina D. *Introduction to Social Macrodynamics: Secular Cycles and Millennial Trends.* Moscow: URSS, 2006; Andrey Korotayev. TheWorld System urbanization dynamics. History & Mathematics: Historical Dynamics and Development of Complex Societies. Edited by Peter Turchin, Leonid Grinin, Andrey Korotayev, and Victor C. de Munck. Moscow: KomKniga, 2006. ISBN 5-484-01002-0. P. 44-62

^ Korotayev A., Malkov A., Khaltourina D. *Introduction to Social Macrodynamics: Secular Cycles and Millennial Trends.* Moscow: URSS, 2006.

http://www.technologyreview.com/news/428434/the-avatar-economy/?ref=rss

http://www.foxnews.com/tech/slideshow/2012/09/26/robots-replacing-humans-everywhere/?intcmp=trending#slide=16

Kurzweil, Ray. "The Singularity." KurzweilAI.net. Web. 3 Dec 2009. <http:>.</http:>

Baase, Sara. A Gift of Fire. 3rd. ed. Upper Saddle River, New Jersey: Pearson Prentice
Hall, 2008. Print.

Kushner, David. "When Man & Machine Merge." Rollingstone. 19 02 2009. Web. 3 Dec
2009.http://www.rollingstone.com/news/story/25939914/when_man_machine_
merge/4>.
http://www.kurzweilai.net/how-do-you-find-the-motivation-to-live-forever

http://elm.washcoll.edu/index.php/2012/09/human-enhancements-are-mankinds-future/

http://www.fastcompany.com/3001739/ibms-watson-learning-its-way-saving-lives

Goertzel, Ben. A Samsung Robot In Every Home By 2020? 26 Mar. 2010. http://hplusmagazine.com/2010/03/26/samsung-robot-every-home-2020/

http://www.kurzweilai.net/video-dial-a-doctor-seen-easing-shortage-in-rural-u-s?fb_action_ids=468392989860360&fb_action_types=og.likes&fb_source=timeline_og&action_object_map=%7B%22468392989860360%22%3A

397719333616060%7D&action_type_map=%7B%22468392989860360%22%3A%22og.likes%22%7D&action_ref_map=%5B%5D

http://tech.fortune.cnn.com/2012/12/04/technology-doctors-khosla/

http://www.latimes.com/health/future/la-he-future-of-hospitals-20120913,0,779637.story

http://blogs.scientificamerican.com/observations/2012/10/24/need-a-hug-baxter-the-human-friendly-robot-debuts-at-m-i-t/?WT_mc_id=SA_DD_20121025

http://spectrum.ieee.org/automaton/robotics/military-robots/darpa-wants-to-give-soldiers-robot-surrogates-avatar-style

Tamar, Lewin. Students Rush to Web Classes, but Profits May Be Much Later
http://www.nytimes.com/2013/01/07/education/massive-open-online-courses-prove-popular-if-not-lucrative-yet.html?pagewanted=all&_r=0

http://www.nytimes.com/2011/01/05/health/05gene.html?pagewanted=2&_r=2&ref=hospitals&

http://www.kurzweilai.net/what-our-civilization-needs-is-a-billion-year-plan

http://www.dailymail.co.uk/sciencetech/article-2041106/Urbee-The-worlds-printed-car-rolling-3D-printing-presses-.html

http://www.smartplanet.com/blog/bulletin/automation-will-soon-touch-every-job-on-the-planet-pre

http://www.forbes.com/sites/timworstall/2012/12/10/that-robot-economy-and-the-rentier-class/diction/968

Dickey, Megan. How Startups Should Prepare For The Day When Technology Merges With Our Brains. 6 Feb. 2013
http://www.businessinsider.com/amazon-is-a-singularity-focused-company-2013-2#ixzz2KFR8JcVb

Brin, David. Singularities and Nightmares: Extremes of Optimism and Pessimism About

    the Human Future. 09 Mar. 2013
http://lifeboat.com/ex/singularities.and.nightmares

Saga 3

Do androids dream of solar power?

http://singularityhub.com/2012/09/24/surging-solar-in-2011-proof-of-ray-kurzweils-bold-prediction/

http://blogs.scientificamerican.com/guest-blog/2011/03/16/smaller-cheaper-faster-does-moores-law-apply-to-solar-cells/

http://www.smartplanet.com/blog/intelligent-energy/graphic-of-the-day-solars-dramatic-rise/17454?tag=main%3Briver

http://www.greencarreports.com/news/1077216_tesla-ceo-elon-musk-half-of-new-cars-will-be-electric-in-15-20-years

Patel, Prachi. "New Battery Could be just what the grid ordered." *Technology Review*. N.p., n. d. Web. Web. 2 Oct. 2012. <http://www.technologyreview.com/news/425565/new-battery-could-be-just-what-the-grid-ordered/>.

http://solarroadways.com/numbers.shtml

http://www.care2.com/causes/renewable-energy-a-strategy-for-long-term-survival.html#ixzz28gPEx64i

http://www.technologyreview.com/featuredstory/429529/how-solar-based-microgrids-could-bring-power-to-millions/

www.ingramcontent.com/pod-product-compliance
Lightning Source LLC
Chambersburg PA
CBHW041241200526
45159CB00028B/8